T0223773

Sustainable Desalination and Water Reuse

Synthesis Lectures on Sustainable Development

Editor

Thomas Siller, *Colorado State Universiy*

The ***Synthesis Lectures on Sustainable Development*** series will publish short books related to sustainable development practices relevant to engineers, technologists, managers, educators, and policy makers. The books are organized around the United Nations Sustainable Development Goals for 2015–2030. Design for sustainability along with the economics of sustainable development will be the common themes for each book. Topics to be covered will span all the major engineering disciplines, along with the natural and environmental sciences that contribute to an understanding of sustainable development. The goal of this series is to make theory and research accessible to practitioners working on sustainable development efforts.

Sustainable Desalination and Water Reuse
Eric M.V. Hoek, David Jassby, Richard B. Kaner, Jishan Wu, Jingbo Wang, Yiming Liu, and Unnati Rao
2021

Oil & Gas Produced Water Management
Eric M.V. Hoek, Jingbo Wang, Tony D. Hancock, Arian Edalat, Subir Bhattacharjee, and David Jassby
2021

Just Technology: The Quest for Cultural, Economic, Environmental, and Technical Sustainability
Thomas J. Siller and Gearold Johnson
2018

© Springer Nature Switzerland AG 2022

Reprint of original edition © Morgan & Claypool 2021

All rights reserved. No part of this publication may be reproduced, stored in a retrieval system, or transmitted in any form or by any means—electronic, mechanical, photocopy, recording, or any other except for brief quotations in printed reviews, without the prior permission of the publisher.

Sustainable Desalination and Water Reuse

Eric M.V. Hoek, David Jassby, Richard B. Kaner, Jishan Wu, Jingbo Wang, Yiming Liu, and Unnati Rao

ISBN: 978-3-031-79507-7   paperback
ISBN: 978-3-031-79508-4   ebook
ISBN: 978-3-031-79509-1   hardcover

DOI 10.1007/978-3-031-79508-4

A Publication in the Springer series
SYNTHESIS LECTURES ON SUSTAINABLE DEVELOPMENT

Lecture #3
Series Editor: Thomas Siller, *Colorado State Universiy*
Series ISSN
Print 2637-7675   Electronic 2637-7691

Sustainable Desalination and Water Reuse

Editors

Eric M.V. Hoek
University of California, Los Angeles (UCLA)
California NanoSystems Institute, CA

David Jassby
University of California, Los Angeles (UCLA)
California NanoSystems Institute, CA

Richard B. Kaner
California NanoSystems Institute, CA
University of California, Los Angeles (UCLA)

Authors

Jishan Wu
University of California, Los Angeles (UCLA)

Jingbo Wang
University of California, Los Angeles (UCLA)

Yiming Liu
University of California, Los Angeles (UCLA)

Unnati Rao
University of California, Los Angeles (UCLA)

SYNTHESIS LECTURES ON SUSTAINABLE DEVELOPMENT #3

ABSTRACT

Over the past half century, reverse osmosis (RO) has grown from a nascent niche technology into the most versatile and effective desalination and advanced water treatment technology available. However, there remain certain challenges for improving the cost-effectiveness and sustainability of RO desalination plants in various applications. In low-pressure RO applications, both capital (CAPEX) and operating (OPEX) costs are largely influenced by product water recovery, which is typically limited by mineral scale formation. In seawater applications, recovery tends to be limited by the salinity limits on brine discharge and cost is dominated by energy demand. The combination of water scarcity and sustainability imperatives, in many locations, is driving system designs towards minimal and zero liquid discharge (M/ZLD) for inland brackish water, municipal and industrial wastewaters, and even seawater desalination. Herein, we review the basic principles of RO processes, the state-of-the-art for RO membranes, modules and system designs as well as methods for concentrating and treating brines to achieve MLD/ZLD, resource recovery and renewable energy powered desalination systems. Throughout, we provide examples of installations employing conventional and some novel approaches towards high recovery RO in a range of applications from brackish groundwater desalination to oil and gas produced water treatment and seawater desalination.

KEYWORDS

reverse osmosis, nanofiltration, desalination, water recycling & reuse, high-recovery, brackish groundwater, seawater, zero liquid discharge, minimal liquid discharge, renewable energy

Contents

CHAPTER 1

Introduction

In the past half century, explosive population growth, rapid industrialization and the trend toward urbanization has lead to over pumping of groundwater for agricultural irrigation and surface water contamination by untreated industrial waste discharge. Combining those factors with upcoming "mega" droughts (> 40 years) due to climate change will leave many areas of the world increasingly vulnerable to water shortages. For example, in 2018, the world counted down to "Day 0" in Cape Town, South Africa—the day when municipal water supplies were largely switched off and residents had to queue up for their daily ration of water. The City of Cape Town was the first major city in the world to "run out" of water, but it likely will not be the last.

The need for more reliable and draught-resilient water supply drives interest in developing a broader portfolio of water supply sources, including traditional fresh ground and surface waters along with non-traditional, more saline water sources such as brackish groundwater, seawater and wastewaters of industrial, agricultural and municipal origins. The higher dissolved solids content of these waters all requires some form of *desalination* to produce water of acceptable quality. The relatively poor quality of non-traditional waters necessitates extensive physical and chemical pre-treatments (pre-disinfection, coagulation, pH adjustment, softening, filtration) and post-treatments (advanced oxidation, stabilization, post-disinfection). This leads to concerns over the increased energy demand of supplying fresh water from non-traditional waters.

According to a report produced by the Energy Producers Research Institute (EPRI) [1], in the early 2000s, about 4–5% of the U.S.'s electricity use went toward moving and treating water and wastewater, while approximately 80% of municipal water processing and distribution costs was for electricity. In addition, the EPRI report states that electricity availability is not a major impediment to economic development or water production, but water is a key constraint on economic development and water limits the potential for new electricity production [1]. Add to this paradigm the rapid growth over the past twenty years in energy-intense desalination and water reuse plants being built to expand fresh water supplies, and we see there is an intrinsic link between energy and water—producing one now, more than ever, depends on the availability of the other. The link between energy and water further encompasses agriculture and food production, which consumes the vast majority of fresh water globally, and, which demands significant energy due to energy-intense fertilizer production along with refrigeration and transportation. This leads us to the *"Food-Energy-Water Nexus."*

The food-energy-water nexus is critical factor in major socio-economic planning decisions and is particularly critical in arid and semi-arid regions where food, water and energy availability are co-limiting development factors. Furthermore, viewing the world through this lens makes clear *our current approaches for supplying food, electricity and water are unsustainable—and they all hinge on the availability of fresh water.* What's worse, since nearly everything remains powered by fossil fuels, the supply of fresh water increasingly becomes a driver of climate change as well as a victim. Additionally, present water treatment does not account for the specific water quality requirements of each end use, which potentially wastes energy by over-treatment; obvious examples include potable water being used for landscape irrigation, industrial process water and toilet flushing. Absent a paradigm shift, the continued urbanization of the world's growing population will expand the energy demands for treatment and transportation of water.

Over the past half century, reverse osmosis (RO) has grown from a niche technology into the most versatile, effective and widely deployed technology for desalination and water reuse. However, there remain certain challenges for improving the sustainability of RO based desalination and advanced water treatment. In low-pressure brackish RO applications, capital (CAPEX), energy and operating (OPEX) costs are governed by product water recovery, which is typically limited by scale formation. In seawater applications, recovery tends to be limited by the salinity limits on brine discharge, while CAPEX and OPEX are driven by electricity used to power the high-pressure feed pumps. The combination of water scarcity and sustainability imperatives, in certain locations, is driving system designs toward minimal and zero liquid discharge (M/ZLD) for inland brackish water, municipal and industrial wastewaters and even seawater desalination.

Herein, we review the basic principles of RO technology, the state-of-the-art for RO membrane, module and system designs, energy and cost-drivers, methods for concentrating and treating brines to achieve MLD, ZLD, and resource recovery, and renewably powered RO desalination systems. We provide both theoretical foundations and applied, industrial insights. This book is intended to be both (1) a basic primer for students and anyone interested in RO technology and applications and (2) a comprehensive reference for those actively engaged in RO technology research, development, engineering, design and operation.

CHAPTER 2

Basic Principles

2.1   OSMOTIC PRESSURE

Osmosis is the process where water molecules move from a solution with high chemical potential (low solute concentration) to another solution of lower chemical potential (higher solute concentration) through a semipermeable membrane. The membrane allows the solvent (water) to pass through, but not solutes (salts). The difference in chemical potential between two solutions can also be described as a difference in osmotic pressure [2]. Accordingly, the osmotic pressure, π, is described by

$$\Delta \pi_m = P_1 - P_2 = \frac{RT\,(\ln a_{i,2} - \ln a_{i,1})}{\overline{V}_i}, \tag{2.1}$$

where $\overline{V}_i$ is the molar volume of water (i), P_1 and P_2 are hydraulic pressures on the membrane side and permeate side, $a_{i,1}$ and $a_{i,2}$ activities of water (i) at the membrane and permeate side, T is temperature, and R is the universal gas constant (8.314 Pa-L/mole-K) [3]. For a dilute solution, the van't Hoff equation is often used to describe osmotic pressure [3]:

$$\pi = \left(\frac{zRT}{M}\right)c, \tag{2.2}$$

where z is the number of ions comprising the salt (e.g., 2 for NaCl, 3 for Na_2SO_4, etc.), c is solute mass concentration, and M is the molecular weight of the solute.

2.2   MINIMUM WORK OF SEPARATION (ΔG_{min})

For salt solutions, the minimum work of separation (ΔG_{min}) is proportional to the osmotic pressure of the solution and is the energy input to overcome the osmotic pressure difference of the solutions in a reversible process. ΔG_{min} is described as [2]

$$\Delta G_{min} = \Delta \pi_m \overline{V}_i = (P_1 - P_2)\overline{V}_i = \frac{RT\,(\ln a_{i,2} - \ln a_{i,1})}{\overline{V}_i}\overline{V}_i = RT \ln\left(\frac{a_{i,2}}{a_{i,1}}\right). \tag{2.3}$$

From Eq. (2.3), one can calculate for a solution of 32 g/l NaCl and seawater, the osmotic pressure and minimum work of separation are approximately 2,792 kPa (400 psi) and 0.8 kWh/m^3 and 2,535 kPa (370 psi) and 0.7 kWh/m^3, respectively. No technology can produce a drop of fresh water from these solutions with lower specific energy demand.

2.3   SOLUTION-DIFFUSION MODEL

The solution-diffusion (SD) model is the most widely applied model to describe the transport through RO membranes. The SD model estimates the local fluxes of the solvent (water) flux (J_w) and solute flux (J_s) through a dense, semi-permeable membrane according to [3]

$$J_w = A\left(\Delta p_m - \Delta \pi_m\right), \tag{2.4}$$

and

$$J_s = B\Delta c_m. \tag{2.5}$$

Here, $A(= D_w c_w K_w / lRT)$ indicates the pure water permeability of the membrane, D_w is the diffusivity of water in the membrane, K_w is the solubility of water in the membrane, l is the membrane thickness, Δp_m the trans-membrane hydraulic pressure, $\Delta \pi_m$ the trans-membrane osmotic pressure, $B(= D_s K_s / l)$ the solute permeability of the membrane, D_s is the diffusivity of the solute in the membrane, K_s is the solubility of the solute in the membrane, and $\Delta c_m(= c_m - c_p)$ the trans-membrane concentration gradient. Here c_m and c_p are solute concentrations at the membrane-solution interface on the feed and permeate sides of the membrane, and $r_s(= 1 - c_p/c_m)$ is the intrinsic local salt rejection.

According to Mulder [2], one can expand and rearrange Eq. (2.5) to obtain $c_p = c_m B/(J_w + B)$, which indicates that local permeate concentration varies with local solute concentration at the membrane surface, solute permeability, and permeate flux. The expression for permeate concentration can be combined with the definition of rejection to produce a useful expression for the local (real) rejection by a RO membrane,

$$r_s = 1 - B/(J_w + B) = J_w/(J_w + B). \tag{2.6}$$

Here, we see the dependence of rejection on the water flux and solute permeability; hence, rejection is not a "membrane property," but rather, a combination of membrane properties (B) and operating conditions (Jw).

Locally, the trans-membrane pressure (Δp_m) is the available hydraulic pressure difference across the membrane, which is described by

$$\Delta p_m = \Delta p_0 - \Delta \pi_m - \Delta p_x - \Delta p_f, \tag{2.7}$$

where Δp_0 is the applied feed pressure, Δp_x is the hydraulic pressure drop along the length of the system, and Δp_f is the additional hydraulic resistance imposed locally by adhered fouling and/or scaling materials. Therefore, local separation performance is influenced by all system variables such as available hydraulic pressure, retentate crossflow velocity and retentate concentration. These system variables are, in turn, influenced by local transport through the spiral wound element as well as membrane fouling and scaling. Ultimately, solute rejection, water flux, osmotic pressure, hydraulic pressure, solute concentration, and retentate velocity can vary significantly from feed inlet to concentrate outlet.

2.4   RECOVERY AND CONCENTRATION FACTOR

Two key performance indicators of any RO system are recovery and concentration factor [3]. Recovery (Y) is the ratio of the total permeate (Q_p) and feed (Q_f) flow rates,

$$Y = \frac{Q_p}{Q_f}. \tag{2.8}$$

Concentration factor (CF) is the ratio of salt concentration in the retentate over that in the feed ($= c_L/c_o$). Taking the overall system product water recovery and observed rejection into consideration yields [4]

$$CF = \frac{c_p}{c_f} = \frac{1 - Y(1 - R)}{1 - Y}. \tag{2.9}$$

Hence, for a solute that is completely ($R = 100\%$) rejected by a membrane, $CF = 1/(1 - Y)$, and for a solute that completely passes a membrane ($R = 0\%$), $CF = 1$. Assuming complete rejection and 50% recovery, the minimum work of separation for solutions of 32 g/l NaCl and seawater, the osmotic pressure and minimum work of separation are approximately 5,584 kPa (800 psi) and 1.6 kWh/m^3 and 5,070 kPa (740 psi), and 1.4 kWh/m^3, respectively.

2.5   THE "THERMODYNAMIC RESTRICTION"

In a full-scale RO process, the feed solution flows through a channel and a portion permeates through the membrane along the length. Thus, the salt concentration increases with increasing length or water recovery, and the available hydraulic pressure drops due to frictional losses from water flowing around the plastic mesh spacers used to create the cross-flow channel. However, the recovery will stop increasing after thermodynamic equilibrium is reached, whereby the solution osmotic pressure equals the feed pressure. This is known as the "thermodynamic restriction" and is illustrated in Figure 2.1 [5].

The water flux is governed by the difference between the trans-membrane hydraulic and osmotic pressures. When the osmotic pressure equals the hydraulic pressure, the net driving pressure becomes zero, which means no more water will permeate through the membrane. At the point where flux is zero, a thermodynamic equilibrium is reached; this is the maximum system length (or recovery) that can be achieved for a given applied hydraulic pressure [5]. Further, assuming 100% rejection, one can deduce a simplified relationship between feed solution osmotic pressure, recovery and the minimum applied hydraulic pressure as,

$$\Delta p = \frac{1}{1 - Y} \Delta \pi. \tag{2.10}$$

Equation (2.10) makes clear that RO system recovery can never reach 100% because as Y approaches 100% the required applied hydraulic pressure becomes infinite.

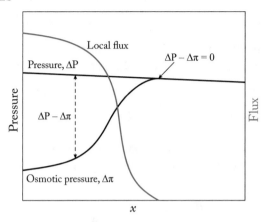

Figure 2.1: Schematic of the thermodynamic limit (reproduced from [5]).

2.6   CONCENTRATION POLARIZATION

In RO separations, permeate flux is initially high, but drops within seconds to a value lower than the initial flux due to concentration polarization (CP). The permeate flux remains steady after the rapid initial decline. This steady-state flux is generally known as the "pseudo steady-state flux" because a more gradual lowering of the permeate flux can occur over a period of days to weeks due to physical membrane compaction (loss of intrinsic water permeability) and weeks to months due to fouling (additional hydraulic resistance on top of that of the membrane), but this will be discussed in more detail below [3]. Generally, in full-scale RO applications, CP makes small contributions to flux decline for clean membranes [4, 7], but can strongly impact flux when enhanced by fouling [8].

Convection from the water flux carries solutes toward the membrane surface from the bulk flowing feed stream, and as solutes are rejected their concentration grows. The ensuing concentration gradient drives diffusive solute flux through the membrane and from the membrane surface back out into the bulk of the flowing feed solution. The competition between these three fluxes defines the elevated steady-state solute concentration at the membrane surface [7], which determines the rate of solute passage and in combination with the local permeate concentration the trans-membrane osmotic pressure. The steady-state, one dimensional solute mass balance in the boundary layer is expressed as [2, 11]:

$$Jc + D\frac{dc}{dx} = Jc_P,\qquad(2.11)$$

where J is water flux, c is salt concentration, c_p is salt permeate concentration, x is the distance away from and perpendicular to the membrane surface, and D is the solute diffusivity.

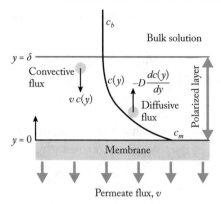

Figure 2.2: Mechanism of concentration polarization (reprinted with permission from [7]).

Integrating Eq. (2.11) with boundary conditions ($x = 0$, $c = c_m$; $x = \delta$, $c = c_b$) gives,

$$J = \frac{D}{\delta} \ln \left(\frac{c_m - c_P}{c_b - c_P} \right), \tag{2.12}$$

where δ is the thickness of boundary layer, c_b is bulk solute concentration [2, 11]. Equation (2.12) can be rearranged into,

$$CP = \frac{c_m}{c_b} = \left[1 - r_s + r_s \exp \left(-\frac{J_w}{k_s} \right) \right]^{-1}, \tag{2.13}$$

where CP is the "concentration polarization" modulus and $k_s (= D/\delta)$ is solute "mass transfer coefficient" [3, 13]. Concentration polarization can be significantly reduced, but never eliminated, by employing higher cross-flow velocities to reduce the boundary layer thickness [14] (Figure 2.2). A detailed description of boundary layer theory is beyond the scope of this book, but we offer an array of mass transfer correlations for different flow regimes and cross-flow channel geometries with and without mesh spacers in Table 2.1.

In addition to increasing salt passage and trans-membrane osmotic pressure, CP enhances fouling and scaling rates by increasing the concentration of fouling and scaling materials on the membrane surface. In turn, back-diffusion of rejected solutes can be hindered by the presence of a colloidal deposit, bacterial biofilm or mineral scale layer [6, 8–10]. These phenomena are variously described as cake-enhanced, biofilm-enhanced and scale-enhanced CP, which leads to enhanced trans-membrane osmotic pressure, solute passage and flux decline.

2.7 REPRESENTATIVE SALINE WATER QUALITIES

Herein, brackish water refers to waters with higher salinity than fresh water (> 500 mg/l), but not with as much salinity as seawater $< 30,000$ mg/l). The salt concentration is generally described by the total suspended solids (TDS) concentration. Seawater generally has TDS

Table 2.1: Mass transfer correlations compiled from various literature sources

$Sh = 1.62\left(\text{Re } Sc\dfrac{d_h}{L}\right)^{0.33}$	Laminar flow in a Tube
$Sh = 0.04\text{Re}^{0.75}\, Sc^{0.33}$	Turbulent flow in a Tube
$Sh = 1.85\left(\text{Re } Sc\dfrac{d_h}{L}\right)^{0.33}$	Laminar flow in a rectangular channel
$Sh = 0.04\text{Re}^{0.75}\, Sc^{0.33}$	Turbulent flow in a rectangular channel
$Sh = \dfrac{kr}{D} = 0.285\left(\dfrac{v}{D}\right)^{0.35}\left(\dfrac{wr^2}{v}\right)^{0.55}$	Laminar ($8000 < \text{Re} < 32000$) flow in a stirred batch cell
$Sh = 0.04\text{Re}^{0.75}\, Sc^{0.33}$	Turbulent flow in a stirred batch cell ($\text{Re} > 32000$)
$Sh_{spacer} = \dfrac{kd_H}{D} = 0.065\ \text{Re}^{0.875}\, Sc^{0.25}$	$\text{Re} < 1000$; commercial RO spiral wound elements

Table 2.2: Definition of saline water types

TDS Concentration (mg/L)	Classification	Typical Water Recovery	Source
<500	Fresh		Rivers; lakes
500–1,500	Low brackish		Municipal wastewater; industrial wastewater; some groundwater
1,500–5,000	Medium brackish	50–90%	Brackish groundwater; agricultural drainage water; industrial wastewater
5,000–25,000	High brackish		Conventional enhanced oil recovery (EOR) oil and gas produced water
25,000–45,000	Seawater		Oceans and seas
>50,000	Brine	30–60%	Oil and gas produced water; geothermal brines

concentrations in the range of 30,000–40,000 mg/l, while brines range from 60,000 to over 300,000 mg/l [15]. Table 2.2 details the different categories of fresh, brackish, seawater and brine.

2.8 MEMBRANE FOULING

Knowledge of the various types of membrane foulants, the mechanisms governing their accumulation on the membrane surface, and their impacts on flux, rejection and differential pressure as well as their prevention through pre-treatment and mitigation through physical-chemical cleaning—all ultimately, drive product water quality, water recovery, energy, CapEx and OpEx of RO membrane installations. There are three distinct types of RO membrane fouling: (1) organic and inorganic colloidal cake formation, (2) biofilm formation, and (3) mineral scale formation [16]. Herein, the types of fouling are categorized this way because they are defined by three distinct mechanisms—deposition (colloidal fouling), bio-growth (biofouling) and precipitation (mineral scaling). In contrast, the associated performance impacts and optimal approaches to prevention and mitigation can be multifold for each fouling type. Generally, colloidal and biofouling cause flux to decline, differential pressure to increase and salt passage to increase, whereas scaling tends to cause the flux to decline and the differential pressure to increase, while limiting the overall product water recovery.

2.8.1 ORGANIC AND INORGANIC COLLOIDAL CAKE FORMATION

Nature & Origin: Nearly all modern RO plants use some type of pre-filtration in the form of 1–10 micron-pore sized cartridge filters; hence, the type of inert particulate materials that foul RO membranes are limited to colloidal particles and organic macromolecules. Colloidal materials range in size from a few nanometers to a few micrometers [17]. Colloidal materials originate from weathering and geochemical processes in soils and sediments (e.g., silicates, clays, iron hydrous oxides, humic and fulvic acids, sulfides, etc.) as well as microbiological processes (e.g., whole algae, bacteria and virus particles, fibrils, cellular exudates, cell debris, etc.) [18]. A unique characteristic of colloidal materials is their metastable state of dispersion (i.e., they don't rapidly settle or float), which is a function of their Brownian nature along with their surface charge and background solution chemistry (ionic strength, pH and presence of multivalent ions), vs. truly dissolved solutes, which remain dissolved in nearly any background solution chemistry [18]. We note the exceptions of certain metal hydroxides, carbonates and silicates, which can be highly sensitive to changes in pH and/or the presence of other multivalent cations; these are not colloidal, but sparingly soluble and can precipitate to form colloidal and larger particles [19].

Indicators & Impacts: Key water quality indicators of colloidal fouling include TOC, TSS, and turbidity, but other diagnostic tests have been developed to assess the RO membrane fouling propensity of feed water. The Silt Density Index (SDI) and the Modified Fouling Index (MFI 0.45) are two popular methods [20]. Membrane suppliers recommend SDI [15] values < 3 of the RO feedwater to avoid increased pressure loss in the modules and to minimize fouling [17]. Modified indexes such as SDI$^+$ also exist; however, they are not sensitive to differences in membrane resistance and testing condition parameters [15, 19, 23]. The relationship between SDI and added hydraulic resistance to permeation appears exponential based on the data shown in

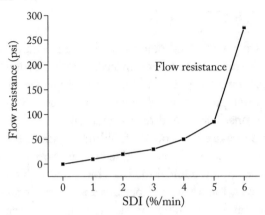

Figure 2.3: Relationship between SDI and flow resistance (reproduced from [13]).

Figure 2.3 [13]. Colloidal fouling causes the flux to decline, the differential pressure to increase and the salt passage to increase. Flux decline can be due to a combination of increased hydraulic resistance to permeation in addition to increased trans-membrane through "cake-enhanced CP," which can also increase salt passage [8]. Increase in system differential pressure occurs because the growing deposit on the surface of the membrane decreases the feed flow channel height and causes blockages in mesh feed spacers—both increase axial hydraulic losses due to friction [4].

Prevention & Mitigation: Colloidal fouling occurs in proportion to the colloid concentration in the RO feed water and the balance between permeate convection toward the membrane and back-migration away from the membrane, which derives from a combination of Brownian diffusion, cross-flow velocity (inertial lift, shear-induced diffusion) and interfacial forces (van der Waals, acid-base, electrostatic) [24–26]. While the best approach to mitigate colloidal fouling is the complete removal by granular media or low-pressure membrane filtration (often in combination with chemical coagulation), another approach is to employ higher cross-flow [13]. High cross-flow can be effective at preventing deposition of particles larger than a few micrometers [25, 26], but it has very little impact on macromolecular natural organic matter (NOM) or nano-colloids due to their small size and Brownian nature. Moreover, these materials tend to internally foul and pass most microfiltration (MF) and ultrafiltration (UF) membranes. To remove such compounds, one would need to employ nanofiltration (NF) or granular activated carbon (GAC) filtration. More details are provided in Chapters 3 and 4.

2.8.2   BIOLOGICAL FOULING

Nature & Origin: Biofouling is a problem in RO membrane systems because polyamide RO membranes rapidly degrade when exposed to free chlorine, ozone, peroxide and other oxidizing biocides, and hence, such biocides cannot be continuously dosed to mitigate bacterial bio-growth. In most RO installations, chlorine is dosed at the head of the plant to limit bio-growth throughout intake and pre-treatment equipment, but just ahead of the RO membranes chlorine is reduced to chloride by addition of a reducing agent like sodium bisulfite. Bacteria may be present in spiral wound RO membrane modules at the point of manufacture because they are not made in completely sterile environments, or bacteria from upstream in the plant may deposit on the RO membranes after they are installed and in operation. Whenever assimilable organic carbon (AOC) is available in the feed water, bacteria grow, reproduce, spread throughout the RO system and release exudates, which causes the flux to decrease, the differential pressure to increase and, sometimes, the salt passage to increase [27–29]. Most historical research—both laboratory and field scale—has focused on seawater RO desalination membrane plants with some emphasis on wastewater RO reuse plants. Not much has been reported in the open literature on brackish groundwater RO systems, but biofouling can be a significant problem there as well [11]. More recently, combined chlorine (chloramine) has been successfully employed to eliminate biofouling in ultra-pure, wastewater and brackish water RO membrane plants; however, seawater RO membranes remain susceptible to chloramine degradation, so biofouling remains a problem.

Indicators & Impacts: The key water quality indicators of biofouling include TOC, AOC and chlorophyll content (particularly in seawater due to algae blooms); typically, RO feed waters are not high in bacterial content. Once seeded, bacteria colonized within a biofilm secrete compounds generally described as extracellular polymeric substances (EPS). It is the EPS that encapsulates phytoplankton and bacterial cells in biofilms that cause flux decline [30]. Biofouling degrades product water quality through "biofilm-enhanced concentration polarization" (BECP), which increases salt passage and elevates trans-membrane osmotic pressure [31]. Moreover, biofouling increases both trans-membrane pressure and feed-to-concentrate differential pressure, thus, increasing the overall cost of water treatment by demanding more pre-treatment, increasing RO system operating pressure (i.e., energy demand) and increasing membrane cleaning frequency, which in turn reduces membrane useful life.

Prevention & Mitigation: Some level of pre-filtration can reduce the load of colloidal and particulate matter that directly foul RO membranes through cake formation and enhanced concentration polarization [8], and also the influent load of viable biofilm forming microorganisms. An effective chemical disinfection regime that inhibits bio-growth throughout the RO system greatly improves performance, reliability and economics by reducing the feed pressure, cleaning frequency, cleaning chemical costs, plant downtime and operator intervention [32]. Biocides such as chloramine and chlorine dioxide have been used to inhibit biogrowth on RO membranes in potable water applications, and in non-potable applications (e.g., boiler feed) non-ozidizing disinfectants such as isothiazolin, chloro-sulfamate, 2,2-dibromo-3-nitrilopropionamide (DBNPA) and stabilized hypobromite are utilized [11]. The ultimate mit-

Table 2.3: Types of salt precipitates in brackish water

Minerals	Carbonate (CO_3^{2-})	Sulfate (SO_4^{2-})	Silica/Silicate (SiO_2/SiO_3^{2-})	Hydroxide (OH^-)	Phosphate (PO_4^{3-})
Most common pairing ions	Ca^{2+}	Ba^{2+}, Ca^{2+}, Sr^{2+}	Fe^{3+}, Al^{3+}, Na^+, K^+, Mg^{2+}, Ca^{2+}	Fe^{3+}, Al^{3+}, Ca^{2+}, Mg^{2+}, Mn^{2+}	Fe^{3+}, Al^{3+}, Ca^{2+}

igation is to shut the system down and chemically clean the membranes to remove any biofilm from the membranes. More details are provided in Chapters 3 and 4.

2.8.3   MINERAL SCALE FORMATION

Nature & Origin: The most common types of minerals that scale RO membranes include carbonates, sulfates, silicates, hydroxides, and phosphates (Table 2.3). Minerals that are of the most concern in brackish water vary depending on the water quality. Among all these minerals, calcite [33] and gypsum [34] are most frequently encountered because of the prevalence of calcium, carbonate and sulfate ions in groundwater, but silica can also be a problem in certain geographies (i.e., Western U.S., Japan) [15, 35]. Therefore, it is important to understand the specific water composition to differentiate the scaling potential of any given water.

Indicators & Impacts: The key feed water quality indicators of mineral scaling are the concentrations of sparingly soluble minerals relative to their solubility product (K_{sp}) [38]. Solubility products of most common scale forming minerals are provided in Table 2.4. In RO systems, the concentration of rejected salts is highest near the exit of the final stage of RO membranes—as determined from the CF—and even higher near the membrane surface due to CP. Hence, the limiting recovery is determined from the first mineral to exceed the product of,

$$K_{sp} \times CF \times CP, \tag{2.14}$$

near the exit of an RO system [4]. Mineral scaling often limits RO product water recovery below the thermodynamic restriction described above [15, 39]. Some minerals tend to precipitate in the bulk and deposit on RO membranes, which manifests like colloidal fouling, but most minerals nucleate and grow a scale directly on the membrane surface. Such scales seal off the covered area of the membrane surface causing rapid flux decline and an increase in differential pressure.

Prevention & Mitigation: The primary approach to preventing scale formation on RO membranes is to limit product water recovery to below the level at which scaling occurs [37]. If that recovery does not meet design targets, the pH can be lowered to inhibit carbonate scaling and/or anti-scalant chemicals can be dosed ahead of RO membranes. The purpose of anti-scalant chemicals is to permit the mineral to exist at super-saturated concentrations without precipitating in the bulk or forming a scale directly on the membrane surface [40]. If acid and anti-scalant

Table 2.4: Solubility products of common minerals at 25°C [36, 37]

Scale	Solubility Product Ksp
$CaCO_3$	2.8×10^{-9}
$CaSO_4$	4.93×10^{-5}
$Ca_3(PO_4)_2$	2.07×10^{-29}
CaF_2	5.30×10^{-9}
$CaSO_3$	6.8×10^{-8}
$CaHPO_4$	1×10^{-7}
$CaSiO_3$	2.5×10^{-8}
$FeCO_3$	3.2×10^{-11}
$Fe(OH)_2$	8.0×10^{-16}
FeS	6×10^{-19}
$Fe(OH)_3$	4×10^{-38}
$MgCO_3$	3.5×10^{-8}
MgF_2	3.7×10^{-8}
$Mg(OH)_2$	1.8×10^{-11}
$Mg_3(PO_4)_2$	1×10^{-25}
$Al(OH)_3$	1.3×10^{-33}
$BaSO_3$	8×10^{-7}
$BaSO_4$	1.1×10^{-10}
$SrSO_4$	3.8×10^{-7}

addition fails to enable a RO plant to meet its recovery goals, additional physical-chemical pre-treatments may be installed ahead of the RO unit (e.g., chemical softening, ion exchange, etc.) and/or brine treatment may be installed to extract more fresh water and further concentrate the RO brine before discharging or disposing of it. More details are provided in following chapters.

CHAPTER 3

State-of-the-Art RO Membranes & Modules

3.1 RO/NF MEMBRANE MATERIALS

Polyamide thin film composite RO membranes now dominate the RO membrane market because of their low-cost fabrication, broad chemical stability, excellent selectivity and high water permeability. This section will briefly highlight the state-of-the-art membranes (Table 3.1) used for brackish water desalination with corresponding membrane chemistry and performance. For more detailed reviews and historical perspectives the reader is directed to reviews by Peterson et al. [41] and Lee et al. [42].

To date, most commercially available NF/RO membranes are based on polyamide thin-film composite (TFC) materials, but some other chemistries exist like poly-sulfonamides and polyamide-ureas, integrally-skinned cellulosic (cellulose acetate, di-/tri-acetate, and their blends) and sulfonated poly(ether sulfone) (sPES). Integrally-skinned asymmetric cellulosic RO membranes are prepared by non-solvent induced phase separation (NIPS), while the TFC-RO membranes are fabricated by polymerizing a thin, dense polyamide layer directly on top of a porous support such as polysulfone via an *in situ* polycondensation reaction between a di- or tri-functional amine and a di- or tri-functional acid chloride [36].

3.1.1 MEMBRANE SURFACE PROPERTIES AND FOULING

Compared with cellulose membrane, the TFC aromatic polyamide membrane exhibits superior water flux and salt rejection, resistance to pressure compaction, wider operating temperature range and pH range, and higher stability to biological attack [52, 53]. However, one of the obstacles to the widespread use of TFC polyamide RO membrane is fouling [54], which is governed by interactions between the membrane surface and foulants in the feed stream [55]. Membrane surface chemistry directly affects fouling propensity [43, 56–61]. The physicochemical properties of RO membrane surfaces, such as hydrophilicity, roughness and charge, are also major factors influencing membrane fouling [61–67].

Membrane fouling in RO systems is closely related to surface characteristics [62, 63], among which surface roughness and hydrophilicity are considered the two most important factors [4, 39, 61, 62, 65, 71–75]. Membranes with smooth and hydrophilic surfaces demonstrate less fouling tendency than those with rough and hydrophobic surfaces [69]. As shown in Fig-

Table 3.1: Representative state-of-the-art RO membranes

Membrane Type	Product	Manufacture	Active Layer Material	Salt Rejection	Pressure (psi)	Feed Concentration (ppm)	Chlorine Tolerance	Features	Ref
Integrally-skinned	ROGA	Koch	Cellulosics	98.0%	420	2000	1 ppm free chlorine	Maximum allowable continuous concentration of free chlorine or similarly active oxidizing agents such as iodine, bromine and ozone is 1 mg/l free chlorine equivalent.	[44]
	HFW1000	Pentair/ X-flow	PES modified/ PES	MWCO: 1,000 Da	—	—	—	Chlorine-resistant, and superb antifouling behavior, hydrophilic, highly asymmetric/ microporous structure	[45]
Thin film composite	X-20	Trisep	Polyamide-urea	99.5%	225	2000	< 0.1ppm	Low fouling due to residual amino groups on surface; and improved silica rejection	[46]
	Duraslick	GE	Aromatic poly-sulfonamide	98.6%	225	800	500 ppm-hrs, dechlorination recommended	Proprietary middle layer creates extreme smoothness creates low-fouling surface	[47]
	NF270	Dow	Linear aliphatic diamines	97%	70	2000	< 0.1 ppm	Good organic (NOM) removal and partial softening preserves water character	[48]
	FT-30	Dow	Fully aromatic polyamide	99.0%	225	2000	< 0.1 ppm	High rejection at low pressure for various contaminants	[49]
	ESPA	Hydranautics	Crosslinked aromatic polyamide	99.4%	150	1500	< 0.1 ppm	Operating at significantly lower pressures	[43]
Thinfilm nano-composite	LG BW 400 ES	LG	Nano-material-incorporated polyamide	99.6%	150	2000	< 0.1 ppm	Significantly increased membrane permeability	[51]

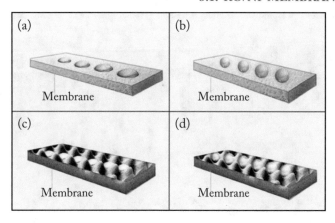

Figure 3.1: Schematic diagram of membrane surface smoothness and hydrophilicity (reprinted with permission from [68]).

ure 3.1, membranes (a) and (c) have hydrophilic surfaces, while membranes (b) and (d) have hydrophobic surfaces. Meanwhile, membranes (a) and (b) have smooth surfaces, while membranes (c) and (d) have rough surfaces. Therefore, membrane (a) is expected to have the best anti-fouling performance, while membrane (d) is expected to be the worst.

Classical definitions of hydrophilicity derive from how a water drop spreads across a material surface: if the droplet beads up into a sphere, the surface is hydrophobic, and if the droplet spreads into a flat lens, the surface is hydrophilic. Generally, more hydrophilic membrane materials tend to be more resistant to fouling, which translates into slower loss of permeability during forward filtration and more complete recovery upon back-washing or cleaning, as shown in Figure 3.2. Free energy of adhesion, ΔG_{132}, which can be calculated through multiple probe liquid contact angle measurements, is a strong indicator of the fouling propensity of any combination of membrane material and foulant material.

Hydrophilic membrane materials attract water so strongly that they require an extreme amount of force (i.e., high flux) to push a foulant particle into intimate contact with a membrane surface. In contrast, hydrophobic membrane materials repel water so strongly that very little force (i.e., no flux) is needed to make foulants stick. Not all hydrophilic materials resist fouling to the same extent. In particular, certain membrane chemistries are macroscopically hydrophilic, but at a molecular scale may attract certain foulant chemistries due to a mixture of acidic and basic moities. For example, polyamide RO membrane coatings possess pendant carboxylate (acid) and amine (base) functional groups. Ultimately, "hydrophilic" does mean "fouling resistant" although there is a strong correlation and often they appear to describe similar fouling behaviors [66].

Hoek et al. [65] developed numerical and analytical models to explain the effects of nanoscale membrane surface roughness features on Derjaguin–Landau–Verwey–Overbeek (DLVO) and extended DLVO interfacial forces. The repulsive interaction energy barrier between a col-

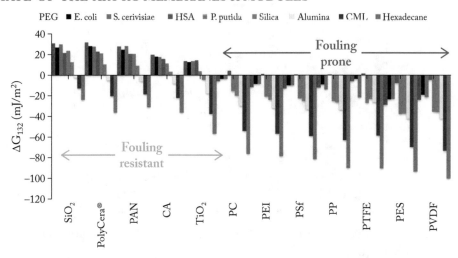

Figure 3.2: Plots of (top) free energy of adhesion or hydrophilicity for common polymeric and ceramic membrane materials and (bottom) free energy of adhesion or fouling propensity for polymeric and ceramic membrane materials along with a wide range of organic, microbial and inorganic fouling materials (reprinted with permission from [73]).

loidal particle and a rough membrane is lower than the corresponding barrier for a smooth membrane. The reduction in the energy barrier is dictated by the radius of curvature of surface roughness features. This work further suggested that the valleys created by the membrane surface roughness produce wells of low interaction energy in which colloidal particles may preferentially deposit.

3.1.2   SURFACE MODIFICATION

There are plenty of ways to alter membrane surface roughness, charge and hydrophilicity via a range of surface modification methods [70, 71] [68]. The methods described below fall into three categories: (1) physical coatings, (2) covalent attachments, and (3) chemical modifications. Each is described in detail with examples from the literature.

3.1.2.1   Physical Coatings

Physical adsorption can be used for modifying and structuring polymer surfaces [79]. Wilbert et al. [73] modified the surfaces of commercial cellulose acetate blends and polyamide RO membranes with a homologous series of polyethylene-oxide surfactants. The results showed that the roughness of a polyamide RO membrane after treatment was reduced, and it exhibited improved antifouling properties in a vegetable broth solution. Besides surfactants, charged polyelectrolytes were also used for surface modification of RO membranes. Zhou et al. [74] modified polyamide RO membranes by electrostatic self-assembly of polyethyleneimine (PEI) on the membrane surface. The application of the PEI layer reversed the charge on the membrane and it was shown to increase the fouling resistance to cationic foulants because of the enhanced electrostatic repulsion as well as increased surface hydrophilicity.

Similarly, Ba and Economy [75] developed a nearly neutrally charged NF membrane by adsorption of a layer of negatively charged sulfonated poly(ether ether ketone) (SPEEK) onto the surface of a positively charged NF membrane. The modified membrane exhibited much better fouling resistance than both the positively and the negatively charged membranes when using bovine serum albumin (BSA), humic acid and sodium alginate as the model foulants. The foulants were less likely to deposit onto the membranes due to the elimination of the charge interaction between the membranes and the foulants.

Surface coating is a convenient and efficient technique for membrane surface modification, and it has been widely adopted to tailor the surface properties of conventional RO membranes. The materials for surface coating are usually hydrophilic polymers containing hydroxyl, carboxyl or ethylene oxide groups. Hachisuka and Ikeda [83] coated hydrophilic and electrically neutral PVA onto polyamide RO membranes and the hydrophilicity of the membrane surfaces was enhanced and the membranes exhibited better antifouling properties after coating with cationic surfactants in the feed. Louie et al. [55, 77] performed another physical coating study of commercial polyamide RO membranes with a very hydrophilic block copolymer of nylon-6 and poly(ethylene glycol). The coated membranes showed reduced surface roughness and exhibited lower flux declines during long-term fouling tests. In addition, Sarkar et al. [85] prepared two types of dendrimer-based coatings for polyamide RO membranes which generally resulted in smooth RO membrane surfaces. The improvement of surface hydrophilicity, the decrease of roughness and the dynamic brush-like topology brought about by this coating all contributed to enhancing antifouling properties. Freeman and co-workers [86–88] developed a series of fouling resistant coating materials by lightly cross-linking, which were used for the surface modification of water filtration membranes including commercial RO membranes. In these methods, the liquid pre-polymer mixture (monomer, crosslinker and photo-initiator) was first coated onto the surfaces of RO membranes and then photopolymerized to form water-insoluble coatings. The testing results indicated that the surface-coated membranes exhibited improved fouling resistance and an improved ability to be cleaned after fouling compared to unmodified membranes.

Tang and co-workers [89–91] fully characterized several widely used commercial RO and NF polyamide membranes by AFM, transmission electron microscopy (TEM), contact angle measurements and streaming potential analysis, and found that some commercial RO membranes were coated with aliphatic polymeric alcohols. The presence of a coating layer could significantly enhance hydrophilicity and reduce surface charge and roughness of membranes, rendering better antifouling properties.

It's worth noting that the coated materials may penetrate into the ridge-and-valley structure of polyamide RO membranes and increase the permeation resistance, resulting in a decline of water flux after modification. Therefore, for practical purposes, the coating layer should have an inherently higher water permeability and be made sufficiently thin to maintain the water permeability. In addition, the modifiers in such a physical modification are only connected with the membrane surfaces via van der Waals attractions, hydrogen bonding or electrostatic interactions, which means the antifouling properties of modified RO membranes may gradually deteriorate due to the loss or leaching of the coating layer during long-term operation.

3.1.2.2   Covalent Attachment

This surface modification technique, including the use of free radical, photochemical, radiation, redox and plasma-induced grafting, is currently used to covalently attach some useful monomers onto the membrane surface which has been covered in reference [52]. Plasma polymerization is a one-step process as the plasma is used to deposit the polymer onto membrane surfaces, while the plasma-induced polymerization utilizes plasma to activate the surface to generate oxide or hydroxide groups, which can then be used in conventional polymerization methods (two-step process) [92]. Compared to the physical modification methods, these modifiers are covalently connected with the membrane surface making them permanent modifications. Therefore, this might be beneficial for long-term operations. However, covalent attachment of specific materials onto RO membranes may require special manufacturing equipment, reagents or complicated operational processes, limiting its practical applications.

Wei et al. [93] performed a radical grafting study with 2,2'-azobis (isobutyramidine) dihydrochloride (AIBA) as the initiator and 3-allyl-5,5-dimethylhydantoin (ADMH) as the grafting monomer. The ADMH-grafted RO membranes showed an increase in surface hydrophilicity and a slight decrease in pure water flux and less adsorption in microbial colonies on the surface. Atkinson and coworkers [94] grafted a bioactive, non-biocidal 2-aminoimidazole (2-AI) to the polyamide active layer of commercial RO/NF membranes, by coupling 2-AI to free carboxylate groups in the active layers using 1-ethyl-3-(3-dimethylaminopropyl) carbodiimide (EDC). Results showed that compared to control (unmodified) membranes, 2-AI-modified membranes inhibited the growth of Pseudomonas aeruginosa biofilms by 61–96%. Yang et al. [95] synthesized a copolymer containing poly-(sulfobetaine) zwitterionic groups, which was covalently grafted on to RO membranes for surface modification. Cell adhesion tests using *E. coli* showed that the modified RO membranes exhibited superior antifouling performance compared to a bare RO membrane. Zou grafted a PEG-like hydrophilic polymer (trimethylene glycol dimethyl ether) onto an aromatic polyamide RO membrane by plasma polymerization. The plasma treated membrane showed enhanced surface hydrophilicity and achieved excellent maintenance of flux compared to the untreated membranes (27% flux reduction). Moreover, the modified membranes exhibited higher flux recovery after cleaning by water. However, the plasma polymerization in their study caused an increased roughness, which is disadvantageous to colloidal fouling [61, 66].

3.1.2.3 Chemical Modification

Hydrophilization can be achieved by chemically modifying a membrane by contact with various water soluble solvents such as acids and alcohols [96, 97]. Mixtures of ethanol and isopropanol and hydrofluoric and hydrochloric acid in water have been used to improve flux and rejection due to the partial hydrolysis and skin modification initiated by the alcohol and acid [98] (Table 3.2).

3.2   MODULE DESIGNS

3.2.1   SPIRAL WOUND ELEMENT

The spiral-wound element (SWE) is the most widely employed module form-factor for RO membranes because it offers a good balance between ease of manufacture, low-cost, stable separation performance, and high packing density [117–119]. The feed spacer strongly influences hydraulic conditions of the feed channel, in particular, the pressure drop vs. cross-flow velocity [110]. The SWE in Figure 3.3 depicts the membrane envelope, a permeate tube, permeate spacers and feed spacers. The feed spacer creates space between the membrane envelopes for feed water to flow and enhances mass transfer near the membrane surface, while incurring additional, but acceptable axial pressure losses [113–116]. However, the feed spacer may cause localized stagnation zones with poor mass transfer that actually encourage fouling and scaling [123]. The axial pressure drop through a membrane module is described by Fanning's equation,

$$\Delta p = 4f \, \frac{L}{d_H} \, \frac{\rho u_0^2}{2},\tag{3.1}$$

where, f is the friction factor (see Table 3.3 for different correlations), L is the module length, ρ is the fluid density, u_0 is the bulk cross-flow velocity, and d_H is the hydraulic diameter. From the various friction factors in Table 3.3, one can deduce that there are different dependencies of pressure drop on the cross-flow velocity for laminar ($\sim u_0$), spacer-filled ($\sim u_0^{1.7}$) and turbulent ($\sim u_0^{1.75}$) flows. It is worth noting that Reynolds numbers typically employed in RO spiral wound elements range from about 100–400, which are strictly laminar. Spiral wound elements are frequently described as employing "turbulence promoting" mesh spacers, but this is not the case; spacers promote undulating flows and recirculating eddies around spacer filaments, but this is not the same as turbulent flows at Reynolds numbers > 2000. Generally, turbulent flows cannot be tolerated by typical commercial RO spiral wound elements because of excessive pressure drops, which could induce telescoping damage and module failure.

3.2.2   SPACER GEOMETRY

One approach toward the design of optimized RO modules is to optimize the feed spacer geometry. Figure 3.4 illustrates different feed spacer configurations that are used in SWE's or are being studied for possible use. The choice on what type of spacer to be used depends on the nature of the feed water. According to a multi-scale modeling study that combined computational

Table 3.2: **Summary of surface modification on RO membranes (*Continues.*)**

Modification Purpose	Coating Agent	Test Conditions	Performance	Ref.
Antifouling	Polydopamine (PDA)	7.5 bar, 500 ppm NaCl, (107–108) cfu/mL bacteria	Biofouling resistance and bacterial adhesion resistance increased	[92]
	Barium chloride (BaCl$_2$)	5 bar, 500 ppm NaCl, 200 ppm BSA	Mineralized membranes had more hydrophilic and negatively charged surfaces. Antifouling resistance improved in addition to flux enhancement (by 1.2 times) and increase in salt rejection (from 96.8% for unmodified membrane to 98.2% for modified membrane).	[93]
	PVA with cationic polyhexamethylene		Modification resulted in higher hydrophilicity, lower roughness, and higher anti-adhesion	[94]
	Sulfonated polyvinyl alcohol (SPVA)	15.5 bar, 2000 ppm NaCl, 2000 ppm foulant (BSA or cetyltrimethylammonium bromide (CTAB))	Modification resulted in increased surface smoothness, hydrophilicity, and electronegativity. With 0.5% (w/v) SVPA, the membrane lost only about 10% of the initial flux after fouling by BSA for 12 h, and the flux recovery reached above 95% after cleaning. Salt rejection increased while water flux decreased.	[95]
	MMA-HPOEM	370 psi, MMA-HPOEM: 0.2 wt-% in a mixture of ethanol and water (1:1 vol. ratio) Seawater: 32,000 ppm & BSA: 30 ppm, seawater & *E. coli*: 7.44 ×10^6 cells/mL	Slower flux decline of modified membrane than the virgin membrane for BSA (37% and 44%) and seawater, antifouling property against reversible and irreversible fouling of *E. coli*, durability of coating layer to chemical cleaning.	[96]
	p(4-VP-co-EGDA) deposition followed by reaction with 3-BPA to obtain zwitterionic pCBAA	Adsorption of BSA and HA, bacterial adhesion of *P. aeruginosa* and *B. licheniformis*.	Fouling resistance in SW feed and HPOEM coating have better fouling resistance under BW feed. Significant enhancement in resisting biopolymers (BSA and HA) adsorption, reduced attachment of seawater bacterial species.	[97]
	P(4-VP-co-DVB) deposition followed by reaction with PS to obtain pyridine-based sulfobetaine	700 psi, NaClO: 1000 ppm for 2, 10, 24 h exposure time, bacterial adhesion of *V. cyclitrophicus*	Coating decreased surface roughness (from 1.3 nm to 0.8 nm), with the 30-nm coating thickness, the water flux is reduced only by ~14% compared to untreated membranes whiteout changing the salt rejection, the addition of 4% DVB produces a major increase in the resistance to chlorine, whereas additions beyond 4% result in minor additional resistance, goo fouling resistance against BSA and bacterial attachment.	[98]
	L-DOPA	18, 50 bar, BSA adhesion, BSA: 100 mg/L, AAS: 100 mg/L DTAB: 50 mg/L	No change in zeta potential after modification, reduction of CA from 55 to 15, intact salt rejection intact (about 97%), increased water flux with increasing the coating duration (1.27 times of original membrane), improved the BSA adhesion resistance, enhanced the organic and surfactant fouling resistance, high water flux recovery ratio (98%).	[99]
	HEMA-co-PFDA	*E. coli* bacterial adhesion, BSA adsorption	By increasing PFDA content, the roughness of surface (<5 nm) and static contact angle increased, coatings were very smooth and conformal, the intermediate chemistry (40% PFDA) showed less alginate adsorption and higher resistance to bacterial adhesion (2 orders of magnitude).	[100]
	ALD-Al2O3	27.6 bar, Bacteria adhesion of *P. aeruginosa*	Coating caused membranes surface tightening, decreasing the roughness, the most hydrophilic surface was obtained with 10 and 50 ALD cycles at temperature 70°C (16 and 27) and 10 ALD cycles at 100°C (27), improved antifouling performance of the membrane, the lowest number of bacteria cells was adhered to surface.	[101]

Table 3.2: (*Continued.*) Summary of surface modification on RO membranes

Modification Purpose	Coating Agent	Test Conditions	Performance	Ref.
Antifouling (cont.)	N-isopropylacryl-amide-coacrylic acid copolymers (P (NIPAm-co-AAc))	5 bar, 500 ppm NaCl, 100 ppm BSA	Modification increased the membrane surface hydrophilicity and surface charge at neutral pH but offer additional resistance to water permeation. Salt permeability decreased and fouling resistance improved.	[102]
	Positively charged antibiotic tobramycin (TOB) and negatively charged poly acrylic acid	15.5 bar, 2000 ppm NaCl, 100 ppm BSA/ sodium alginate (SA)	Under optimized condition, the modified membrane showed 18% increase in water flux and slightly enhanced (0.4%) salt rejection. Also, the modified membrane demonstrated 37% and 26% higher flux than the virgin membrane after three-cycling fouling of BVA and SA solution, respectively.	[103]
	Sericin	5 bar, 500 ppm NaCl, 100 ppm BSA	Modification resulted in increased surface hydrophilicity, enhanced surface negative charge, smoothed surface morphology, and decreased pure water permeability and salt permeability coefficient. Fouling resistance improved.	[104]
	PEI	8 bar, 90 ppm NaCl, 50 ppm DTAB	Pure water permeability decreased by 37%. Salt rejection, fouling resistance, and surface hydrophilicity improved	[74]
Chlorine tolerance	GO	15 bar, 1000 ppm NaCl, 6000 ppm NaOCl	During the first 2 h of chlorine exposure, the salt rejection decreased from 95.3% to 91.6% for the modified membrane while, for the unmodified membrane, salt rejection dropped to 80%. After 16 h of chlorine exposure, the salt rejection for the modified membrane was 75% compared to 63% for the unmodified membrane.	[105]
	Ring-opening by DMAP followed by SPGE and glycerol coating	1.5 MPa, NaClO: containing 100 ppm free Cl_2 for 540, 1620, and 3780 ppm/h Cl_2 exposure time	SPGE membranes were more hydrophilic and smoother, improved chlorine stability, with increasing DMAP concentration, water flux increased and salt rejection decreased, by increasing SPGE concentration water flux decreased and salt rejection decreased (above 0.05 wt%).	[106]
	GA	5 bar, BSA: 100 mg/ L, SDS: 200 mg/ L, DTAB: 10 mg/L	By increasing the content of PVA, the surface roughness increased, CA decreased (from 60.5°C (raw) to 40°C), surface negative charge decreased, pure water flux increased when PVA content is <150 mg/L, salt rejection increased from 98.05% to 98.42%, decreased water flux slightly, good durability of modified membranes, improved antifouling property and chlorine stability.	[107]
	Dimethylamino-ethanol (DMAE)	1.6 MPa	The modified membrane showed a decrease in CA in comparison with the nascent membrane and the tertiaryamine membranes (from 60° to 28°), stronger chlorine resistance and more powerful anti-biofouling properties.	[108]
Scale resistant	Methacrylic acid (MAA)	The active polyamide (PA) layer of a thin-film composite (TFC) synthesized membrane was activated with an atmospheric plasma, followed by a solution free-radical graft polymerization (FRGP) of a water soluble methacrylic acid (MAA) monomer, at 60°C and initial monomer concentration of 5–20% (v/v), onto the surface of the PA-TFC membrane.	Structuring of the PA-TFC membrane at equivalent FRGP reaction conditions resulted in membranes of higher permeability (by a factor of 1.3–2.26) relative to a commercial RO membrane of a similar surface roughness (~70 nm) and salt rejection. Flux tests of membrane mineral scaling demonstrated that membrane mineral scaling propensity can be measurably reduced, relative to commercial membranes of the same salt rejection, while yielding equivalent or higher water permeability. The onset time for gypsum scaling for the optimal membrane surface (prepared at 10% (v/v) initial MAA concentration) was retarded by a factor of 2–5 relative to the commercial RO membrane.	[116]

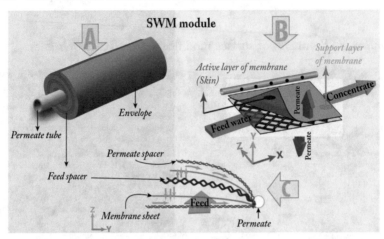

Figure 3.3: A schematic view of a spiral wound membrane module out of the pressure vessel (a), unwrapped with only two envelopes (b), and a side view of the feed channel (c) (reprinted with permission from [124]).

Table 3.3: Pressure drop correlations for open and spacer-filled cross-flow channels

	Laminar	Turbulent
Tubular (open, no mesh)	$f = 16 \cdot Re^{-1}$	$f = 0.079 \cdot Re^{-1/4}$
Channel (open, no mesh)	$f = 24 \cdot Re^{-1}$	$f = 0.133 \cdot Re^{-1/4}$
Spacer-filled channel	$f = 6.23 \cdot Re^{-0.3}$	N/A

fluid dynamics simulations of spacer filled channels with a full-scale RO process model, the woven mesh spacers employed in most RO SWE's are well-suited for high-salinity, low-recovery applications where osmotic pressure dominates and short system designs limit the axial pressure losses [7]. However, the opposite is true for low-salinity, high-recovery systems where axial pressure losses dominate and osmotic pressure is not significant. Ladder-type or more open, corrugated spacers are more suitable for low salinity applications to minimize hydraulic pressure losses through the length of typical multi-stage systems designs. The lower resistance of the ladder-type spacers also makes them more suitable to be cleaned with air-flow, by which higher shear rates help remove the particles and biofilms [118, 125].

3.2.3   NOVEL SPACERS

There are a large number of studies that have adopted different approaches to develop novel membrane spacers for water treatment applications. Alteration of feed spacer geometry is a potential option to reduce the deleterious impacts of hydraulic pressure loss, CP and fouling. Good

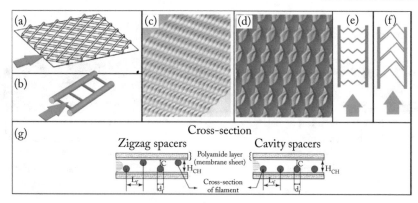

Figure 3.4: Typical spacer design; rhombus diamond-type (a), ladder-type (b), monolayer helix (c), double layer helix (d), zigzag-type (corrugated) (e), and herringbone spacer (f). Two possible cross-sections: zigzag-type and cavity-type cross-sections (g). The possible geometrical parameters are adaptive height c/H_{CH}, and aspect ratio (L_f/h). L_f is the distance between the two filaments, H_{CH} is the channel height, d_f is the average diameter of the filaments, and c is the gap between the membrane and the cross filament (reprinted with permission from [118, 125]).

spacer configurations should reduce the build-up of fouling deposits and decrease CP while not creating axial pressure losses [126, 127]. Rejected species accumulation can be suppressed by enhancing back-mixing from the boundary layer adjacent to membrane to the liquid bulk [128]. Several strategies have been adopted to modify feed spacers such as by surface coating [129], altered geometry design [130] or three-dimensionally printed feed spacers [131] and use of electrically conductive spacers [125] as summarized in Table 3.4.

Three-dimensional (3D) Printed Spacer Technology™, the patented spacer technology of AquaMembranes, is proven to significantly reduce the pressure drop, increase the membrane surface area in SWE's, boost permeate flow and reduce membrane fouling [126]. Unlike traditional membrane spacers that trap particles and biofilms within the spacer mesh creating a high pressure drop, the innovative 3D-printed spacers can optimize flow patterns and mixing through the SWE. This optimization leads to lower energy consumption, reduced system footprint, improved cleaning and potentially longer element life. Because the spacer is printed directly on the membrane surface, the spacer has a less constrictive geometry than a conventional mesh spacer material. This opens a pathway to reduce the total energy consumption along with allowing 30–50% more membrane area in the same diameter of SWE [126].

Table 3.4: Summary of feed spacer modifications (*Continues.*)

Parameter modified	Specification	Effects			Ref
		Hydraulic pressure loss	Flux	Fouling	
Surface coating	Functionalization of PP films by a metal chelating feed spacer charged with copper ions	—	—	Copper ions were employed to disinfect water from *E. coli* bacterial cells with a concentration of 3.0×10^5 cells/ml	[134]
	Coating the membrane and feed spacer with nano silver particles	—	Higher flux	Silver-coated spacer showed antimicrobial activity	[135]
	Copper coated spacer	No remarkable changes	No noticeable change	Lower biofouling only in initial stages	[129]
	Silver, copper and gold coated feed spacers	Reduced pressure loss	—	Biofouling is not prevented	[130]
	Polydopamine and polydopamine-g-PEG coated feed spacers and membranes	—	Lower flux reduction	To anti-fouling bovine serum albumin, and *Pseudomonas aeruginosa*	[128]
	Commercial polypropylene feed-spacer modification via sono-chemical deposition of zinc oxide nanoparticles	—	Lower flux reduction	To hinder biofilm layer formation	[131]
	Commercial polypropylene feed spacer functionalization via sono-chemical deposition	—	Improved flux	To control biofouling development	[122]
	Polymer brushes	No remarkable changes	No noticeable changes	Not effective in terms of surface biomass accumulation	[129]
Filament cross-section	Round filaments compare to other shapes	Lower pressure drop	Lower flux	—	[139]

Table 3.4: (*Continued.*) Summary of feed spacer modifications

Parameter modified	Specification	Effects			Ref
		Hydraulic pressure loss	Flux	Fouling	
Filament torsion	—	—	Flux enhancement due to destabilization of the boundary layer	—	[133]
Hydrodynamic angle in UF (230 < Re < 1661)	UF-90°	Highest pressure drop	Highest flux	—	[134]
	UF-45°		16–25% lower flux due to channeling	—	
Orientation of feed spacer	45° versus 90°	45° had a higher pressure drop	45° had a better mass transfer	At 90° orientation, particles are deposited in a transverse band across the entire spacer cell, and in the normal orientation, deposition occurred primarily in the center of the cell	[119]
Spacer thickness	Increase from 0.7 to 0.86 mm	Lower pressure drop in thicker spacer	—	Lower fouling in thicker spacer	[142]
Number of layers	Increased number of layers	Higher turbulence and pressure drop	Higher flux	—	[143]
Aspect ratio	Increase of aspect ratio	Lower pressure drop	—	—	[144]
Porosity	Increase of porosity	Minor decrease in pressure drop	Minor decrease in flux (2–10%)	—	[134]

CHAPTER 4

System Design & Performance

4.1 SYSTEM DESIGN

4.1.1 SINGLE STAGE AND MULTI-STAGE ARRAYS

The basic unit of an RO system is the SWE. A number of SWEs (between 5 and (8) are housed in a single "housing" or "pressure vessel." A collection of pressure vessels in parallel forms a stage, and the number of stages in series forms an array. A tapered array is used when the feed flow rate is reduced significantly between stages due to recovered permeate. In such a cascading array configuration, the number of parallel pressure vessels is reduced in each stage. Cascading arrays sometimes have multiple recycle and bypass streams.

In single-stage arrays, the feed is pumped into the RO module at a designated pressure, which then splits into a permeate stream and a concentrate stream. The permeate stream is the product water and the concentrate stream is considered "industrial waste" [145]. In a multi-stage setup, the brine from one stage is sent as the feed to the next stage, which enables higher overall recovery. In some designs, an inter-stage booster pump may be required to increase hydraulic pressure above the osmotic pressure of the concentrate at the end of the final stage. Use of inter-stage booster pumps can save energy in exchange for higher CapEx. Single-stage and multi-stage RO systems designs are depicted in Figure 4.1.

The theoretical minimum energy required for desalination is equal to the energy needed to bring the feed water to the osmotic pressure of fluid permeating across the membrane (represented by the area under the osmotic pressure curve in Figure 4.1a). In practice, to ensure that water permeates across the membrane at all points along the membrane channel, the pressure applied to the feed volume must be at least equal to the osmotic pressure of the concentrate stream exiting the membrane channel ($P_H = \Pi_c$); the energy is represented by the area of the rectangle of height P_H. The difference between these two areas represents the energy that must be used to operate with a single stage. Staged membrane operations can save RO desalination energy costs. The first stage operates at a lower applied pressure $P_{H,1}$ and percent recovery. The concentrate from the first stage is then pressurized by a booster pump to $P_{H,2} = \Pi_c$ to achieve the desired percent recovery. This configuration needs to bring a smaller volume up to $P_{H,2} = \Pi_c$, thus saving the energy represented by the smaller hatched rectangle.

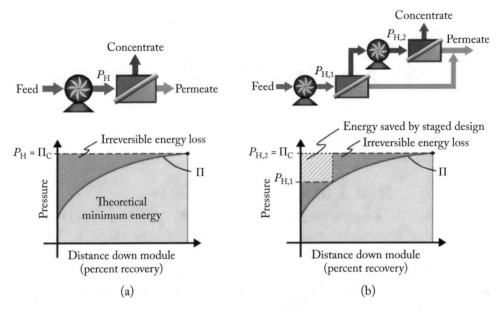

Figure 4.1: Energy use of single-stage and multi-stage RO systems (reprinted with permission from [139]).

4.1.2   MULTI-PASS RO SYSTEM DESIGN FOR HIGH PURITY

In seawater and brackish groundwaters, boron is most likely to exist as boric acid, which is uncharged, poorly hydrated and very small in size. Boron rejection by seawater RO (SWRO) membranes is over 90% for many commercial SWRO products and seawater boron levels are fairly low (3–6 ppm) [147]. However, boron rejection by brackish water RO (BWRO) membranes is considerably lower (55–65%) and boron concentrations can be in excess of 50 ppm [148]. However, at pH values above 11, $B(OH)_4^-$ dominates and its rejection is very high [149]. However, operating at such high pH increases carbonate and hydroxide levels, which can induce scaling particularly from hard ions (e.g., $CaCO_{3(s)}$, $Mg(OH)_{2(s)}$) [150]. Multi-pass RO systems allow for the removal of bulk TDS including nearly all hard ions in a first pass with the pH adjusted down to inhibit carbonate and hydroxide scaling. Then the pH is elevated prior to a second pass targeting near complete removal of boron from the low TDS permeate. In the production of ultra-pure water, it is difficult to reject gases such as carbon dioxide and ammonia. A triple pass system has been developed for the removal of such gases. In this method, a gas permeable hydrophobic membrane is provided between the two RO systems [147].

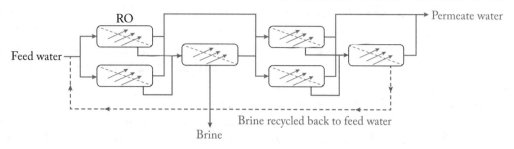

Figure 4.2: A schematic of a double pass RO system with each pass comprising a two-stage array.

4.2   KEY PERFORMANCE INDICATORS

Given the design of full-scale RO systems with multiple elements in series inside a pressure vessel with potentially multiple passes and multiple stages of pressure vessels in cascading arrays, tracking the local performance throughout the length of a system can be difficult. Companies that manufacture RO membranes offer process models for full-scale system design, but the manufacturers do not reveal the math and physical principles on which they are based. In the open literature, a physics based model for full-scale RO systems that accounts for local pressure, flow and solute balances throughout was validated against full-scale plant data with excellent agreement [4]. The basic principles of this open source model are provided below.

4.2.1   PRESSURE AND FLUX

The trans-membrane pressure (Δp_m or "TMP") is the available hydraulic pressure difference across the membrane. In a full-scale RO system, the TMP changes from inlet to outlet. The two main drivers for change in TMP from inlet to outlet are the decreasing available hydraulic pressure due to axial frictional losses and the increasing in osmotic pressure due to water permeation along the length of the system. Additional TMP losses occur due to fouling and scaling, which increase trans-membrane hydraulic resistance and axial hydraulic losses. TMP is described by

$$\Delta p_m = \Delta p_0 - \Delta \pi_m - \Delta p_x - \Delta p_f, \tag{4.1}$$

where Δp_0 is the applied feed pressure, $\Delta \pi_m$ is the trans-membrane osmotic pressure, Δp_x is the hydraulic pressure drop along the length of the system and Δp_f is the hydraulic resistance of the local fouling layer. In a full-scale plant, due to the high costs of pumping, the trans-membrane pressure is an important contributor to the overall economic viability of the RO system.

The differential pressure drop (or "delta P"), Δp_x, is the difference between the pressure of the feed stream at the inlet of a membrane housing and the pressure of the concentrate stream at the outlet. The differential pressure should remain constant as long as the flow and temperature are steady and provided there is no fouling or scaling. Hence, it is an important indicator of system performance. In multi-stage systems, an increase in differential pressure at the first stage could be due to organic/colloidal or biological fouling; if there is an increase in differential pressure in the last stage, it suggests scaling may be happening. Both fouling and scaling can also increase the trans-membrane hydraulic pressure drop—locally and globally. Again, in multi-stage systems where TMP is monitored for each stage, increase in TMP in the first stage indicates fouling and in the last stage indicates scaling.

From the change in TMP, it is clear that the flux will change (decrease) from inlet to outlet of a full scale RO system. The system average flux $(\bar{J})$ is the total permeate flow divided by the total membrane area; however, when considering the local flux over the length of the entire system, it can be determined directly by integrating the local permeate flow over the system length,

$$\bar{J} = \frac{\int_0^L J_w W dx}{WL}. \tag{4.2}$$

Here W is the flow channel ("leaf") width and L is the system length. It should be noted that water flux varies with feed water temperature, T. In practice, it is often corrected using,

$$J_T = J_{25}(TCF), \tag{4.3}$$

where TCF is the temperature correction factor, which is determined from,

$$TCF = (1.03)^{T-25}, \tag{4.4}$$

where T is in units of degrees Celsius. Another empirical, but more robust temperature correction method relies on testing membrane water flux and salt rejection at different temperatures and employing an Arrhenius-type equation such as,

$$A = A_{25} \exp\left[K_{wT}\left(\frac{1}{273.15 + T} - \frac{1}{298.15}\right)\right], \tag{4.5}$$

$$B = B_{25} \exp\left[K_{sT}\left(\frac{1}{273.15 + T} - \frac{1}{298.15}\right)\right], \tag{4.6}$$

where A_{25} and B_{25} are the water and solute permeability coefficients at 25°C. Also, K_{wT} and K_{sT} are the temperature coefficients for water and solute transport, which are obtained by regression analysis of water and solute permeability data obtained at different temperatures [152].

4.2.2 PRODUCT WATER QUALITY

In RO systems, the key performance indicator is permeate water quality, which for a given compound can be determined from the system average permeate concentration $(\bar{c}_p)$

$$\bar{c}_p = \frac{\int_0^L J_w c_p W dx}{\int_0^L J_w W dx} \tag{4.7}$$

or by the observed rejection,

$$R = 1 - \frac{\overline{c}_p}{c_f}. \tag{4.8}$$

Here, W is the width of a leaf in a spiral wound element, c_f is the feed concentration, and, x, is the location along the total system length, L. The TDS should be measured at the outlet of each pressure vessel so as to isolate the problem if there is a water quality drop detected. A universal drop in water quality will be indicated by broad membrane damage. If permeate quality degrades in only one stage, the problem could be local membrane fouling. If it is only observed in an individual pressure vessel, there could be a damaged or defective membrane as well as o-ring damage in a single or multiple brine seals [159].

4.2.3 PRODUCT WATER RECOVERY

Product water is recovered along the length of a full-scale RO system as water permeates through the membrane. The overall system recovery is determined by integrating the local permeate production over the length of the system as

$$Y = \frac{\int_0^L J_w W dx}{Q_0}, \tag{4.9}$$

where Q_0 is the influent flow rate. So, in conventional RO systems, product water recovery increases with water flux and system length.

4.2.4 SPECIFIC ENERGY CONSUMPTION

Energy consumption by RO processes is the product of the applied pressure and the feed flow rate divided by the electrical efficiency of the pump used to generate the pressure and flow. Specific energy consumption (SEC) is the energy per unit volume of permeate, and is expressed by,

$$SEC = \frac{\Delta P_f Q_0}{\eta_P Q_P} = \frac{\Delta P_f}{\eta_P Y}, \tag{4.10}$$

where η_p is the pump efficiency. Various methods to reduce SEC have been identified [154]:

1. Using highly permeable membranes to reduce feed pressure [155]

2. Using an energy recovery device (ERD) [156]

3. Operating at higher recovery [157]

4. Using renewable energy sources to subsidize the electrical energy demand [158]

5. Optimizing RO configurations and operating conditions using mathematical models [159]

When an ERD is employed, the high pressure brine passes through a device that transfers the pressure to an equivalent flow rate of the low pressure feed, thus reducing the duty of the primary pump [160–162]. The specific energy recoverable (SER) by an ERD is,

$$SER = \frac{Q_b P_b}{Q_P} \eta_{ERD} = \left(\frac{1}{Y} - 1 \right) P_b \eta_{ERD}, \qquad (4.11)$$

where η_{ERD} is the pressure transfer efficiency of ERD, where P_b is the brine pressure. However, to compute the net specific energy demand, SEC_{net} ($\neq SEC - SER$) one must consider that there is a feed pump and a booster pump each pressurizing a different portion of the flow by a different amount. The SEC_{net} comes from an energy balance on the feed and booster pumps ($= \Delta P_{fp} \cdot Q_{fp}/\eta_{fp} + \Delta P_{bp} \cdot Q_{bp}/\eta_{bp}$), which upon normalizing by Q_p and re-arranging becomes,

$$SEC_{net} = \Delta P_{fp}/\eta_{fp} + \left(\Delta P_{fp} - \Delta P_{bp} \cdot \eta_{ERD} \right) \cdot (1/Y - 1)/\eta_{bp}, \qquad (4.12)$$

where subscripts fp and bp denote the feed pump and booster pump, respectively.

The specific energy consumption for seawater desalination RO plants is between 3.7–5.3 kWh/m^3 [163]. The use of high rejection membranes in a two pass system led to a specific energy consumption of 3.6–3.7 kWh/m^3 [164]. For small size plants (capacity less than 100 m^3/day), SEC is much higher, around 10 kWh/m^3 [165]. In modern, high-efficiency SWRO plants, SEC_{net} values as low as 1.80–2.20 kWh/m^3 can be obtained. One example of such a system is present in Sal Island-Cape Verde for a SWRO plant of capacity 1,000 m^3/day [158]. The affordable desalination coalition (ADC) demonstrated SWRO with a capacity of 200–300 m^3/day at an energy cost of 1.58 kWh/m^3 using Filmtec® membranes at a flux of 6 L/m$^2 \cdot$ h [166].

4.3   CAPEX AND OPEX DRIVERS

On September 22, 1961, John F. Kennedy was quoted to have said [163], "If we could ever competitively, at a cheap rate, get fresh water from saltwater, this would be in the long-range interests of humanity which could dwarf any other scientific accomplishments" [167]. Today, we can get fresh water from saline waters, but the cost remains a limiting factor at times. Table 4.1 provides a summary of the costs of desalination for low-salinity (TDS = 500 $\sim$ 2,500 mg/l) and high-salinity (TDS = 2,500 $\sim$ 10,000 mg/l) brackish water desalination plants. The cost summary presented in this table is based on comparative analysis of over 40 brackish water desalination plants worldwide [164]. Table 4.2 presents the range of water production costs of medium (i.e., having a production capacity of 10,000–100,000 m^3/day) and large size (i.e., having fresh water production capacity of over 100,000 m^3/day) seawater RO desalination projects. Information for this table is compiled based on a comparative review of over 50 desalination projects in the United States, Australia, Europe, the Middle East, the Caribbean, and other parts of the world. As seen in Table 4.2, at present (in 2018 US$) the average industry-wide

Table 4.1: Water production costs of medium and large size BWRO desalination plants [168]

Classification	Cost of Water (US$/m³)	
	Low-Salinity BWRO Plants	High-Salinity BWRO Plants
Low-end bracket	0.2–0.4	0.3–0.6
Medium range	0.5–0.8	0.7–1.0
High-end bracket	1.0–1.5	1.3–1.8
Average	0.7	0.9

Table 4.2: Water production costs of medium and large size SWRO desalination plants [168]

Classification	Cost of Water (US$/m³)
Low-end bracket	0.5-0.8
Medium range	0.9-1.5
High-end bracket	1.6-3.0
Average	1.1

cost of production of fresh water by SWRO is approximately US$1.1/m³. Comparison of Tables 4.1 and 4.2 reveals that on average seawater desalination production costs are 1.2–1.6 times higher than those for producing fresh water by high-salinity and low-salinity brackish water desalination, respectively. When comparing some individual projects however, this difference could be significantly higher [168].

It is interesting to note that the cost difference is not strictly proportional to salinity. Often, low-salinity sources may contain additional contaminants such as silica, cyanide, iron, manganese, or large quantities of organics and dissolved gases that have a profound impact on plant construction costs because their removal usually requires additional treatment steps and expenditures. In addition, both low and high-salinity brackish water plants use the same type of BWRO membrane elements, pressure vessels, and pumps, which have the same unit costs per processed capacity (i.e., their costs are mainly determined by plant production flow, recovery, and are not as significantly impacted by source water salinity as they are by production flow). Figures 4.3 and 4.4 show a breakdown of the water production costs of low-salinity and high-salinity BWRO plants by their main components: direct (construction) and indirect capital costs; and power and other operation and maintenance (O&M) costs. For low-salinity desalination plants, construction costs (i.e., direct capital costs) are typically the largest component of the

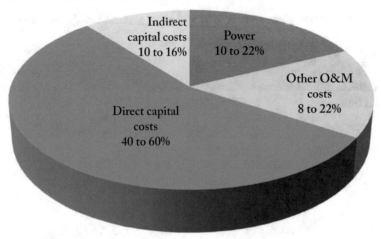

Figure 4.3: Typical costs of water breakdown for low-salinity (TDS = 500 ∼ 2,500 mg/l) BWRO plants (reprinted from [168]).

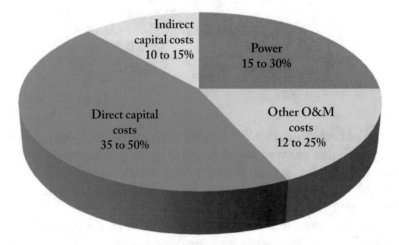

Figure 4.4: Typical costs of water breakdown for high-salinity (TDS = 2,500 ∼ 10,000 mg/l) BWRO plants (reprinted from [168]).

total cost of water production. The wide range of these costs is mainly attributed to the economy of scale, and differences in intake and concentrate disposal cost components.

Comparative analysis of Figures 4.3 and 4.4 indicates that in high-salinity BWRO plants power is a slightly larger proportion and CapEx a slightly lower proportion of the total water production costs because of the higher operating pressure required to overcome the higher feed water osmotic pressure. However, the energy cost component is not directly proportional to the

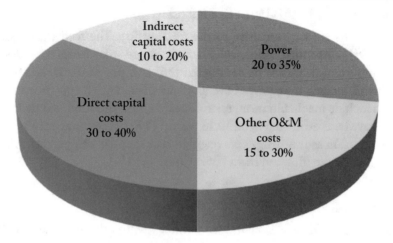

Figure 4.5: Seawater reverse osmosis plant—cost of water breakdown (reprinted from [168]).

salinity because high-salinity BWRO plants operate at lower recovery and apply energy recovery devices, whereas for low-salinity BWRO plants it is not cost-effective to apply energy recovery.

Comparison of Figures 4.3 through 4.5 indicates that capital costs for BWRO facilities are usually a higher portion of the total water production expenditures than those for SWRO plants (45–60% vs. 30–40%). In BWRO projects, the energy contributes 10–30% of the total costs compared to SWRO projects where energy contributes 20–35% of the total, and in extreme conditions for remote plant locations with high unit energy costs, energy expenditures can be considerably higher.

4.3.1   INTAKE AND DISTRIBUTION

Subsurface intakes such as vertical beach wells, infiltration galleries, slant wells and horizontal wells tap into brackish water sources as the intake for BWRO plants. Water sourced from beach wells have been shown to have elevated iron and manganese concentrations that result in high pre-treatment costs. A major contributing factor for intake is the cost of the intake pump. Studies on cost of intake for seawater RO have been shown to be 4–5% of the total capital investment [165]. Water distribution costs can range from a few U.S. cents to a few U.S. dollars/m^3 [168]. The main contributor to the water distribution cost is due to vertical lifting. In Turkey, a 78 km pipeline with a capacity of 75 million m^3 per year would lead to an additional cost of 25–34 cents/m^3 [167]. As per the literature, the cost of water produced by BWRO plants ranges from $0.38 per m^3 to $0.75 per m^3, not including distribution [168]. Thus, the distribution costs contribute 25–50% to the entire cost, depending on the distance of transport.

4.3.2   FEED QUALITY AND PRE-TREATMENT

RO system operations are very sensitive to the feed quality. The feed water salinity, the silt density index (SDI), suspended solids, colloids, oil, grease, metal oxides, scales and organic substances are important factors to be considered [169]. A pretreated feed is much less likely to foul a membrane, leading to improved overall plant efficiency. The pre-treatment techniques that can be used include media filtration, microfiltration, ultrafiltration and nanofiltration. pre-treatment costs have been estimated to be $0.15 per m³ [173]. Cost comparisons between media and membrane filtration as pre-treatments revealed that they are comparable, while media has a lower CapEx, membrane filtration has a lower OpEx [171].

4.3.3   PRODUCT WATER RECOVERY

The quality of the brine depends on the feed water quality, the permeate quality, the pre-treatment methods used and the cleaning procedures applied [176]. In coastal desalination plants, brine is released into the sea and disposal costs are lower. For inland desalination plants, depending on the concentration of the brine, the cost could be considerably higher [177]. The cost of brine disposal in all cases lies between 5–33% of the cost of seawater desalination [178]. The volume of brine to be disposed of has a major impact on cost. Therefore, evaporation in large ponds has been widely used to reduce volume [179].

4.3.4   CHEMICAL CLEANING AND MEMBRANE REPLACEMENT

Cleaning and replacement in one module affects the pressure in the other modules, thus affecting the operation of the RO system [141]. Annual membrane and cartridge filter replacement costs are proportional to the replacement frequency of these consumables, which in turn depends on the source water quality and the plant design [168]. Since often at the conceptual stage of an SWRO desalination project, the number of cartridge filters is unknown, a rule of thumb that can be used for preliminary assessment of the total number of the cartridge filters needed for the plant is: 25 cartridge filters/1,000 m³/day of plant fresh water production capacity. The typical useful life of ultra- and micro-filtration pre-treatment membranes is 3 to 7 years, although some membrane suppliers provide membranes with useful life warrantees for up to 10 years [168]. Therefore, their annual replacement costs range between 10.0% and 33.3% of the initial installed membrane cost.

The useful life of first-pass seawater membranes is typically between 5 and 7 years, while the typical useful life of the second-pass brackish water reverse osmosis (BWRO) membranes is 10 years [168]. As a result, the typical annual average SWRO membrane replacement rate is 14.3–20.0% of their initial installed costs. A rule of thumb that can be applied for determining the total number of SWRO membrane elements in the first pass of the RO system is 80 SWRO elements/1,000 m³/day and the total number of BWRO elements in the second pass of the RO system is 30 BWRO elements/1,000 m³/day [168]. However, the membrane replacement costs

Table 4.3: Unit chemical costs [164]

Chemical	Unit Cost (US$/kg)
Chlorine gas	0.6–1.1
Sodium hypochlorite	2.2–3.5
Ferric sulfate and ferric chloride	0.4–1.2
Sulfuric acid (93% H_2SO_4)	0.06–0.1
Citric acid	1.6–2.5
Biocide	3.0–5.5
Sodium hydroxide (50% NaOH)	0.65–0.85
Sodium bisulfite	0.35–0.55
Antiscalant	1.6–4.0
Ammonium hydroxide	0.6–1.1
Hydrated lime	0.3–0.4
Calcite	0.05–0.08
Carbon dioxide	0.08–0.12
Sodium tripolyphosphate (corrosion inhibitor)	1.6–3.2
Other cleaning chemicals (US$/m³ of permeate)	0.005–0.008

should be determined based on the actual number of membrane elements that will be used for the project.

Chemical costs are highly variable from one location to another and are mainly dependent on the source water quality, the selected pre-treatment processes, and the target product water quality. Table 4.3 presents unit costs for various chemicals frequently used in desalination plants. The actual chemical cost values for a given project have to be established based on quotes from local suppliers of the site-specific chemicals. Costs of chemicals is a variable expenditure and typically is in the range of US$0.030 per m³ to US$0.060 per m³ of product water.

4.3.5   BRINE DISPOSAL

The cost of reject brine disposal depends on the brine characteristics, the level of treatment before disposal, the disposal method and the environment and brine volume [181]. Current disposal methods include sewer discharge, deep-well injection, surface ponds, percolation pits, surface discharge, and ocean discharge.

4.3.5.1   Sewer Discharge

Brine discharge to sewers is a popular method of disposal for brackish water RO plants. In this method, brine from desalination facilities is discharged to a nearby wastewater collection system and sent to a wastewater treatment plant (WWTP) for treatment and final disposal. It is an attractive option because it allows desalination facilities to use the existing wastewater infrastructure, cutting the costs needed to construct their own. The existing wastewater in sewers can also help with transporting brine over long distances, thus alleviating energy requirements in brine conveyance. Costs are lowest when the RO plant is in close proximity to the WWTP. However, this method's feasibility is also highly site specific.

Sewer disposal requires a permit from the local sewage agency, and its viability depends on the capacity of the wastewater collection system and treatment plant. To be used for brine disposal, the wastewater facility must have the volumetric capacity to accommodate wastewater and the additional volume of brine. RO brine would also change the quality and characteristics of the wastewater effluent. For example, the high salinity of concentrate could cause the wastewater components to aggregate differently than they typically would. The addition of RO brine may reduce the biochemical oxygen demand of the sewage and may reduce its temperature as well. The new mixture must remain compliant with the WWTP discharge permit, so treatment of the concentrate may be required before discharge to sewers. It is also worth noting that the high TDS concentration of RO brine can have a negative effect on the biological treatment processes of WWTPs. It could also impact the use of reclaimed water from the WWTP because compounds like sodium, chlorides, boron, and bromides from RO concentrate can have a harmful effect on microorganisms, plants, and soil. Another factor to take into consideration is the fee that the WWTP operators will charge for its use. In some cases, they may require an initial fee to purchase capacity for the brine, then additional operating costs based on the volume and pollutant concentration of the brine. For these reasons, sewer discharge is typically only feasible for small and medium-sized brackish water RO facilities whose brine would not put too much strain on the volumetric capacity of the plant and the chemical composition of the effluent.

4.3.5.2   Surface Water Discharge

Surface water discharge involves the disposal of brine through outfalls to open bodies of water, such as lakes, rivers, and the ocean. Disposal in this way requires a National Pollutant Discharge Elimination System (NPDES) permit to ensure that brine will not impair the receiving waters. In most cases, surface water disposal will not be allowed if the brine will result in a 10% or more increase in the receiving water salinity. If concentrate salinity values are too high, or if harmful components are present, the concentrate may be diluted with freshwater or treated to meet an acceptable water quality. Contamination of the receiving waters is the key environmental concern in this method of brine disposal. The key cost factors in surface water discharge are those of concentrate transport, outfall construction and operation, and concentrate treatment to meet permit requirements and environmental monitoring. Therefore, cost can be very low if the

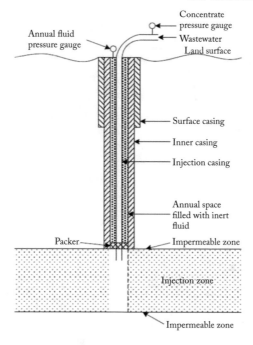

Figure 4.6: Basic structure of an injection well (reprinted with permission from [181]).

desalination plant is located near the body of water and if the brine requires little treatment prior to disposal. It is especially economical when an existing outfall of a wastewater treatment plant or power plant can be used. Surface water discharge is currently the most common method of desalination brine disposal in the U.S. This is reflected in the fact that a large number of desalination facilities are located near bodies of water.

4.3.5.3   Deep-Well Injection

Deep well injection is a disposal method that involves injecting RO brine into deep underground, confined aquifers that are not used for drinking water. The wells are typically composed of concentric pipes encased in concrete, as can be seen in Figure 4.6, and range from 300–2,400 m in depth. Its operation requires a low surface area footprint, has relatively low energy requirements and provides an alternative to long-distance pipelines to distant bodies of surface water. It is a widely used disposal method for inland facilities, but many factors must be considered to determine the feasibility of deep well disposal of RO brine.

4.3.5.4   Evaporation Ponds

Disposal in evaporation ponds is another conventional method of concentrate management for inland desalination facilities. It involves pumping the brine into shallow, lined basins and al-

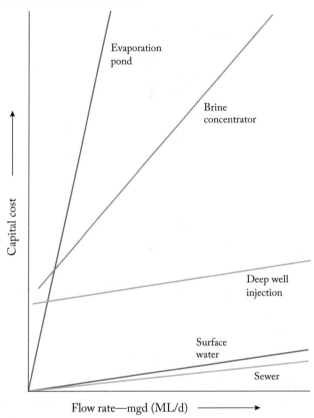

Figure 4.7: The relative capital costs of conventional concentrate disposal methods as a function of concentrate flow rate. Site-specific costs may vary significantly (reprinted with permission from [182]).

lowing water in the brine to evaporate naturally by solar irradiation. When water evaporates, salts from the brine accumulate in the pond and are removed periodically for disposal in land-fills. Often, a series of ponds will be constructed to allow for continuous brine discharge. A key advantage of this method is that it is very simple: it does not require any advanced technology, and little maintenance is required for operation. It is also a viable and proven disposal option for facilities in arid inland areas far removed from surface waters or WWTPs. However, similar to the previously discussed methods, the use of evaporation ponds is site-specific and several factors must be considered before its implementation.

4.3.6   SUSTAINABILITY BENEFITS OF MINIMAL AND ZERO LIQUID DISCHARGE (ZLD/MLD)

As discussed above, the issues and limitations of common brine disposal methods can be briefly summarized as follows. The common brine disposal methods, without exception, are limited by their possible environmental detriments. Thus, rationally, the solution to minimize the environmental impact is to minimize the volume of brine discharged or introduced to the receiving environment, since brine generation is inevitable. In contrast to the limitations of conventional low-recovery desalination brine disposals, the benefits of ZLD/MLD design/operation are:

- More sustainable by maximizing fresh water recovery

- Application is not limited by water quality

- Minimal environmental impact

- Few regulatory limitations

- Minimal practical restrictions (e.g., scale, footprint, etc.)

The primary deterrent of ZLD/MLD is its high cost to afford the additional capital equipment, energy demand and other operating costs. However, if we look deeper into the "real cost" of the brine disposal, the investments on ZLD/MLD can be more reasonable and justified.

For the conventional ZLD/MLD, the additional cost, compared to conventional low recovery RO design/operation, is typically introduced by the brine concentrator and crystallizer, which at the cost of high energy consumption produces more water (TDS: 30–50 mg/L), also called distillate. The overall recovery is thus enhanced via a combination of the RO system with the brine concentrator and crystallizer, which will be further discussed in the brine management section.

Specific energy requirements of the brine concentrator are 60–90 kWh/kgal distillate, whereas the crystallizer requires 180–250 kWh/kgal distillate [183]. These energy consumptions are much higher than the SEC of a normal seawater desalination (10–15 kWh/kgal permeate), which makes it the major contribution to the overall additional cost. However, based on the following discussion, the investments in ZLD/MLD are reasonable.

In theory, SWRO plants aiming at 100% recovery, will have an amortized cost range from $29.12–$59.87, at different capacities, where the large capacity plant tends to have a cheaper rate primarily because the CAPEX ratio on the total cost is much higher in the low capacity plant where not enough water is treated. Whereas the normal cost of a conventional SWRO plant is $3–$8/kgal permeate, this is nearly one order of magnitude cheaper than a ZLD SWRO plant in theory.

This huge gap between the cost of conventional operation and a ZLD design makes the ZLD less feasible in seawater desalination. In addition, typically the SWRO plant will dump the brine into the ocean. If the brine is oversaturated, under ZLD operation, the receiving ecosystems nearby the discharging spot will be heavily impaired. Therefore, the ZLD operation is not

Table 4.4: Limitations of the common brine disposal methods

Strategy	Limitation
Surface water discharge	Environmental permitting Availability of suitable receiving bodies Impact on downstream water supplies
Deep well injection	Environmental permitting Potential for inducing earthquakes
Evaporation ponds	Environmental permitting Available area Capital cost
Sewer discharge	Environmental permitting Site specific Treating capacity

Table 4.5: Theoretical seawater ZLD cost summary [185]

Case	Capacity[1] (MGD)	Capital Cost ($/gpd)	O&M Cost ($/kgal)	Amortized Cost[2] ($/kgal)
ZLD-1	1	$41.11	$19.11	$29.12
ZLD-2	0.5	$56.77	$20.29	$34.25
ZLD-3	0.25	$86.13	$23.67	$44.24
ZLD-4	0.125	$133.11	$28.07	$59.87

Notes:
1. Assumes 100 percent recovery (i.e., ZLD feed flow = ZLD treated water flow)
2. Amortization over 20 years at a 6 percent annual interest rate

practical in SWRO plants. However, it is still meaningful to push the SWRO recovery higher, under the pressure of water scarcity in certain regions. It is not practical to have ZLD in SWRO plants, but the story is different in BWRO plants.

Unlike SWRO plants where the brine is normally discharged into the ocean, the brine generated from inland BWRO plants are the biggest problem that needs to be tackled. As shown in the figure, the biggest portion in OPEX is the brine disposal, up to 51% [185]. This is because it requires not only the cost of disposal, but also that of brine transportation. In certain regions where the sewer discharge or the surface discharge are unavailable, while deep-well injection and land use are also regulated, then the brine disposal cost will increase exponentially, making the total cost abnormally high. In fact, to date, under the demand of sustainable development, policies on brine disposal are becoming more and more strict. One of the most direct and explicit

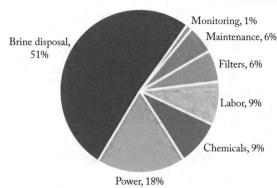

Figure 4.8: Industrial desalination plants OPEX breakdown (reprinted from [185]).

Table 4.6: Theoretical 80% recovery BWRO plant cost summary [186]

| | Capacity (MGD) | | | Capital Cost | O&M Cost | Amortized Cost[1] |
Case	Feed	Permeate	Brine	($/GPD)	($/kgal)	($/kgal)
RO-1	5	4	1	$0.58	$0.30	$0.44
RO-2	2.5	2	0.5	$0.80	$0.40	$0.59
RO-3	1.25	1	0.25	$1.11	$0.60	$0.87
RO-4	0.625	0.5	0.125	$1.54	$0.70	$1.07

Note: 1. Amortization over 20 years at a 6 percent annual interest rate

solutions to this situation is to push the recovery higher, approaching to ZLD/MLD, so that the brine generated is minimized and the cost on it will thus be minimized.

Table 4.6 summarizes the amortized cost of a normal BWRO plant that runs at 80% recovery. If ZLD is coupled with this BWRO operation, the amortized costs will range from $6.18–$12.83, similar to the cost of a seawater desalination plant. For the same plant, if the recovery is pushed higher, from 80–90%, as shown in Table 4.8, the amortized cost will be slightly reduced, since the original cost is also not significant. But the difference appears when ZLD is applied, as shown in Tables 4.9 and 4.10. There is a 39% unit cost reduction from an 80% recovery operation to a 90% recovery operation. It is also noteworthy that the amortized cost, $3.77 at a 5 MGD BWRO plant that runs at 90% recovery, is very attractive. Since it should be noted that these tables, Tables 4.6, 4.7, 4.8, 4.9, 4.10, assume that the OPEX cost is relatively low. However, in real cases, as already mentioned, the OPEX can be abnormally high, thus, the actual cost of a non-ZLD/MLD operation is possibly higher than the ZLD design. Hence, ZLD/MLD desalination is feasible at industrial desalination plants where brine disposal is a huge problem that needs to be tackled. In other words, it is necessary to achieve a high recovery

Table 4.7: Theoretical 80% recovery BWRO+ZLD cost summary [186]

Case	BWRO		ZLD		BWRO+ZLD	
	Permeate Flow (MGD)	Amortized Cost ($/kgal)	Distillate Flow (MGD)	Amortized Cost ($/kgal)	Product (MGD)	Amortized Cost ($/kgal)
RO-1→ZLD-1	4	$0.44	1	$29.12	5	$6.18
RO-2→ZLD-2	2	$0.59	0.5	$34.25	2.5	$7.32
RO-3→ZLD-3	1	$0.87	0.25	$44.24	1.25	$9.54
RO-4→ZLD-4	0.5	$1.07	0.125	$59.87	0.625	$12.83

Table 4.8: Theoretical 90% recovery BWRO cost summary [186]

Case	Capacity (MGD)			Capital Cost1 ($/GPD)	O&M Cost1 ($/kgal)	Amortized Cost2 ($/kgal)
	Feed	Permeate	Brine			
RO-1	5	5	0.5	$0.55	$0.25	$0.38
RO-2	2.5	2.25	0.25	$0.76	$0.35	$0.53
RO-3	1.25	1.125	0.125	$1.05	$0.60	$0.85

Table 4.9: Theoretical 90% recovery BWRO+ZLD cost summary [186]

Case	BWRO		ZLD		BWRO+ZLD	
	Permeate Flow (MGD)	Amortized Cost ($/kgal)	Distillate Flow (MGD)	Amortized Cost ($/kgal)	Product (MGD)	Amortized Cost ($/kgal)
RO-1→ZLD-1	4.5	$0.38	0.5	$34.25	5	$3.77
RO-2→ZLD-2	2	$0.59	0.5	$44.24	2.5	$4.90
RO-3→ZLD-3	1	$0.87	0.25	$59.87	1.25	$6.75

RO desalination, as a solution to the dramatically increasing pressure on the brine management as well as water scarcity and sustainable development.

Table 4.10: Comparison of RO+ZLD cost at 80% vs. 90% RO recovery

Treated Water Capacity (MGD)	80% Recovery Amortized Cost ($/kgal)	90% Recovery Amortized Cost ($/kgal)	Unit Cost Reduction
5	$6.18	$3.77	39%
2.5	$7.32	$4.90	33%
1.25	$9.54	$6.75	29%

CHAPTER 5

Methods for Achieving High Recovery

5.1 PRE-TREATMENT

To mitigate fouling and scaling as well as to optimize their performance, RO systems require effective pre-treatments to supply superior feedwater. Ineffective pre-treatment will lead to severe membrane fouling, higher membrane cleaning frequency, lower recovery, higher operating pressure, poor product quality and short membrane life, ultimately impairing plant productivity and operational costs (Table 5.1). Accordingly, pre-treatment selection and optimization is a key factor for a successful RO desalination system operation [73, 74, 187–193].

5.1.1 SUSPENDED SOLIDS REMOVAL

5.1.1.1 Turbidity and TSS Removal

Water can be categorized according to its salinity level. Three main categories of salinity are seawater which has a total dissolved solids (TDS) concentration of about 35,000 mg/L or more, brackish water or medium-salinity water with a TDS concentration of 1,000–15,000 mg/L and fresh water with a concentration below 500 mg/L [225]. Important parameters in designing a BWRO system are the feed source, the feed concentration, the feed flow, the product flow, and the product quality. A good BWRO system tries to minimize the feed pressure and the number of membranes used, but tries to maximize the permeate quality and recovery [226–230]. Brackish water composition varies widely. Therefore, the composition of it must be known before any design can be carried out. A complete and accurate water analysis must be obtained [225]. Water composition needs to be analyzed. An example brackish water composition is shown in Table 5.2.

Membrane fouling is associated with the existence of particles and colloidal material in the feed water, since they concentrate on the membrane surface. The occurrence of mineral scaling involves complex chemical reactions, which proceed both in solution and at the membrane-water interface. In either case, the kinetics of mineral nucleation and growth depend upon the feed solution chemistry (e.g., chemical composition, pH, and the presence of natural organic matter) [27]. Turbidity is recommended to be less than 0.2 NTU (nephelometric turbidity units) for a successful RO treatment [13]. High turbidity and TSS will lead to severe membrane fouling and membrane damage, eventually impairing the overall recovery as well as the total cost.

Table 5.1: RO pre-treatment techniques and their advantages, disadvantages

Pretreatment techniques	Advantages	Disadvantages	Ref.
Chlorination	Effective disinfection; Cost-effective; Odor removal	Membrane oxidation and degradation;	[194–199]
Ozonation	Strong oxidation;	Bromate generation; Residuals management	[200]
UV radiation	Cost-effective Easy operation; Short HRT	Performance impaired in light scattering water	[201, 202]
Chemical coagulation	Considerable TSS, turbidity removal	Unstable product quality; Inadequate dosing exacerbates membrane fouling/ scaling; Overdosing introduces Fe^{3+} and Al^{3+} and triggers scaling	[203]
Cartridge and bag filters/ Media filter	Assisting coagulation with TSS, turbidity removal; SDI control	Unstable product quality; Big footprint	[1, 204]
MF	Lower the OPEX; Reduce the pretreatment footprint; Stable product quality; Lower the RO fouling, and the RO replacement/cleaning expenditure	No removal of undesired inorganics; Raised CAPEX; Unable to remove virus; Susceptible to chlorine; Colloid fouling, biofouling	[188, 205–209]
UF	Lower the OPEX; Efficient TSS/NTU removal Lower the RO fouling, and the RO replacement/cleaning expenditure Better effluent quality than MF;	No removal of undesired inorganics; Raised CAPEX; Mineral Scaling, organic fouling;	[188, 205–209]
NF	High effluent quality Significantly reduced RO OPEX	Raised CAPEX; NF membrane mineral scaling; NF cleaning/ replacement	[188, 205–209]
Anti-scalants	Effectively repress mineral scaling	High cost; Overdosing causes biofouling	[210–212]
Lime, lime-soda, high-lime, warm lime, chemical de-silication	Effectively repress mineral scaling	Require filtration process afterwards; pH-shift expensive for highly buffered waters; Sludge dewatering and disposal is expensive	[204]
Ion exchange	Remarkable hardness removal; Less sludge than chemical precipitation; Mitigate RO mineral scaling	Does not reduce concentration below silica saturation without metal addition; pH control required; Brine disposal; IEX membrane reactivation	[213, 214]
Electro-dialysis	No microbial/colloid contaminants removal	High CAPEX; Brine disposal	[215, 216]
Capacitive deionization	Eco-friendly; No brine;	High CAPEX; Electrode replacement	[217]
Compact accelerated precipitation softening	Remarkable hardness removal; Mitigate RO mineral scaling	High CAPEX; No organic removal	[218–223]
Warm lime softening/ sodium aluminate	Repress silica fouling	Overdosing causes high SDI	[224]

Table 5.2: Example of brackish water characteristics [226, 230]

Property	Value	Unit
TDS	1800	mg/L
Turbidity	370	NTU
TSS	300	mg/L
pH	6.7	—
Total hardness ($CaCO_3$)	1146	mg/L
Conductivity	6.35	mS/cm
Silica	50	mg/L

Therefore, turbidity and TSS, as the basic references for feed quality, should be minimized for achieving a high recovery of the BWRO system.

5.1.1.2   Chemical Coagulation

To eliminate suspended and colloidal material, as well as to minimize the turbidity in the feed-water, coagulation/flocculation is one of the most conventional but effective methods, which also contributes to lowering the cost. Coagulation is charge neutralization of finely divided and colloidal impurities in feedwater into masses that can be filtered. In addition, particles have negative electrical charges, which cause them to repel each other and resist adhering together. Coagulation, therefore, involves neutralizing the negative charges and providing a nucleus for the suspended particles to adhere to. Flocculation is the bridging together of coagulated particles [224].

Various coagulants, either organic or inorganic, have been tested and studied. The choice on what coagulant(s) to use depends on the quality and other properties of the feed. MFI values after pre-treatments of $FeCl_3$ flocculation and PAC (powdered activated carbon) adsorption is found to be significantly decreased [231]. $AlCl_3$ and $NaAlO_2$ are employed to remove the phosphate from the feed, with the purpose of mitigating membrane fouling [232]. By applying $AlCl_3$ and $NaAlO_2$ coagulation, the normalized water flux can be increased to 66% and 64%, respectively, upon a total permeation volume of 250 ml. $NaAlO_2$ appears to be more effective in mitigating membrane fouling [232]. However, conventional coagulants, both aluminum and iron based, may result in adverse membrane performance.

In a study of alum and ferric chloride [203], rapid deterioration in specific flux (up to 60% over 100 h of operation), as well as progressive reductions in salt rejection (typically 34% over 500 h of operation) are observed in the test of alum. The foulants caused by residual coagulant are primarily aluminum hydroxide and aluminum silicate materials. In contrast to the performance of alum, which suggests declining membrane flux, specific flux obtained using ferric chloride and chloramines increases over time, whereas salt rejection significantly decreases. Note that RO membranes may be degraded by residual iron catalyzing a chlorine-amide reaction on their membrane surfaces [203].

5.1.1.3   Cartridge and Bag Filters/Media Filter

In conventional RO pre-treatment, a media filter, bag or cartridge filters are commonly combined with chemical coagulation, to remove the flocs generated in coagulation and flocculation as well as to eliminate turbidity and TSS of the feed. Filtration, usually considered a simple mechanical process, involves the mechanisms of adsorption (physical and chemical), straining, sedimentation, interception, diffusion, and inertial compaction [1, 204]. Quartz sand, silica sand, anthracite coal, garnet, magnetite, and other materials may be used as filtration media. Silica sand and anthracite are the most commonly used types [68, 204, 233, 234].

The media type, surface charge, size, and geometry of both the contaminant particles and the media particles affect the efficiency of removing solids [193, 234]. Sharp, angular media form large voids and remove less fine material than rounded media of equivalent size. The media must be coarse enough to allow solids to penetrate the bed for 2–4 in. Although most suspended solids are trapped at the surface or in the first 1–2 in. of bed depth, some penetration is essential to prevent a rapid increase in pressure drop [204]. Periodic washing of filters is necessary for the removal of accumulated solids. Inadequate cleaning permits the formation of permanent clumps, gradually decreasing filter capacity. If fouling is severe, the media must be cleaned chemically or replaced [235]. For cleaning of rapid downflow filters, clean water is forced back up and through the media. In conventional gravity units, the backwash water lifts solids from the bed into wash troughs and carries them to waste.

Depending on the design of the media support structure and the accessory equipment, the backwashing rate differs. A high-rate backwash can expand the media by at least 10%. Backwash rates of 12–15 gpm/ft^2 or higher are common for sand, and rates for anthracite may range from 8–12 gpm/ft^2. Low-rate backwash, with no visible bed expansion, can be combined with air scouring [189, 204]. Clarifier effluents of 2–10 NTU may be improved to 0.1–1.0 NTU by conventional sand filtration. The filter cartridges are usually 1–10 μm [13, 236]. Particles larger than five microns can be removed by a classical method. New media designs of cartridge filtration are being explored, such as electro-adsorptive media, which can achieve removals of 6 log units for *Escherichia coli* [237].

5.1.1.4   Membrane Filtration (MF/UF)

The disadvantages of conventional processes include: (i) fluctuation of the quality of the RO feed caused by changing raw water conditions, (ii) difficulty in maintaining an SDI < 3.0, especially during high turbidity feed water conditions, (iii) low removal of particles smaller than 10–15 microns, and (iv) the coagulant impact on RO membranes [188, 205]. Feed water with low quality produced by conventional pre-treatment, specifically in many cases with consider-

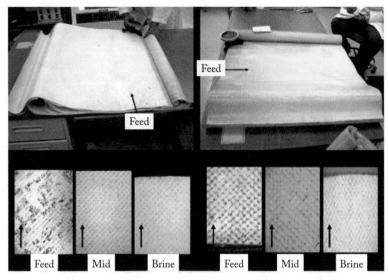

MF/UF pretreatment   Conventional pretreatment

Figure 5.1: RO membranes performance comparison between conventional and MF/UF pre-treatment (reprinted with permission from [243]).

able colloids, suspended particles and hardness [13, 238, 239], challenges the tolerance of RO membranes, the consequences of which are stronger fouling tendencies for the membranes and lower recovery. The cost of membrane cleaning and changing is increased as well. Therefore, the exploration of pre-treatments with higher efficiency is inevitable (Figure 5.1).

Membrane filters represent a "new" trend that is commonly applied in RO pre-treatment nowadays [188, 205–209]. The membrane filter refers to large pore size membranes in MF (0.1–0.35 μm), UF (0.01–0.05 μm) and NF (1–2 nm). Membrane pre-treatment systems have proven to be very effective for turbidity removal as well as the removal of non-soluble and colloidal organics contained in the source water. For brackish water and wastewater feeds, the case for UF/MF pre-treatment is normally accepted. The world's largest industrial market for RO is in China. There, UF/MF is specified for nearly all RO projects irrespective of feed type [240].

Studies suggest that membrane pre-treatment can provide permeate waters with SDI < 2 [238], turbidity < 0.05 NTU, which promises higher recovery for RO, longer RO membrane lifetimes, lower chemical consumption and reduced membrane cleaning frequency leading to reduced downtime [13, 234]. However, more operational energy and costs are required by membrane pre-treatment [188, 207, 240–242]. In the long run, the advantages brought by membrane pre-treatment for brackish water may exceed the overall system cost, whereas for seawater it may not be worth enough due to the high salinity. The additional cost of UF/MF may be paid for simply by the savings on chemicals and consumables. The additional cost of UF

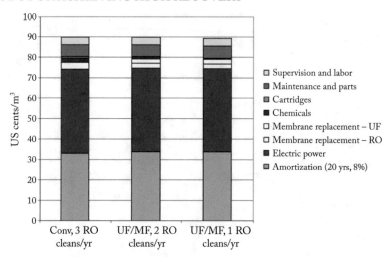

Figure 5.2: Cost comparisons for seawater applications (reprinted with permission from [240]).

in terms of CAPEX and membrane replacement is 2.9 cents/m^3. However, UF reduces RO replacement, saving 1.2 cents/m^3, and reduces chemical costs for both dosing, and RO cleaning. If the RO cleanings are reduced from three cleanings per year to two cleanings per year, the saving amounts to 1.7 cents/m^3, which with the RO replacement saving pays for the UF/MF. If two cleanings are saved, UF/MF becomes cheaper than conventional pre-treatment by 0.7 cents/m^3. This ignores other potential benefits arising from the 33% space saving of UF/MF, and the opportunity to increase RO flux and recovery [240]. Moreover, by saving one RO cleaning per year, the total water cost of a UF/MF pre-treated RO system is equivalent to that of conventional pre-treatment, purely based on the savings in RO replacement costs, chemicals, and cleaning downtime [240]. Figure 5.2 explicitly suggests that higher feed quality can be achieved after MF/UF membrane pre-treatment.

5.1.2 ORGANIC & BIOFOULING INHIBITION

5.1.2.1 Chemical Disinfection

To eliminate the threat of biofouling and organic fouling, various methods can be applied to disinfect the feed water, e.g., UV radiation, ozonation or addition of oxidizing agents, such as chlorine, chlorine dioxide, hydrogen peroxide, chloramines and sodium hypochlorite [172, 194, 244–251]. Chlorine can be added either to control biological growth or to disinfect the UF/MF membranes in the membrane filtration stage. Even though chlorination has some major drawbacks (e.g., hazardous disinfection by-products), it is still the most widely applied disinfectant in conventional water treatment plants due to its ease of operation and low cost. RO desalination systems use intermittent or continuous chlorination with target residual chlorine concentrations

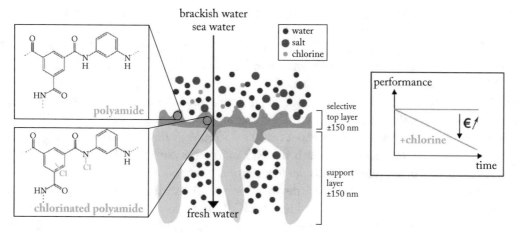

Figure 5.3: Scheme of chlorine degradation on a polyamide membrane (reprinted with permission from [194]).

of 0.2–4 ppm, the contacting time of which varies from 15–30 min [249]. In addition to the adverse effect of producing DBPs, chlorination also jeopardizes the polyamide-based membrane performance, since the residual chlorine induces polyamide degradation [194–199]. When chlorine is present in the feed water, it attacks the PA (polyamide) layer. Membrane chlorination eventually leads to a strong performance decline over time.

The poor chlorine-resistance of PA-membranes is due to the sensitivity of the amidic nitrogen. Two mechanisms of chlorine degradation on polyamide are commonly presented (Figure 5.3). The first mechanism is N-chlorination, which means the oxidizing species (HOCl) attacks the amidic nitrogen via a reversible N-chlorination reaction. First, the end amine groups are chlorinated and subsequently the internal-amide bonds. The N-chlorination reaction has a first-order dependence on free chlorine and is promoted in acidic media [252, 253]. The reverse reaction takes place in alkaline conditions, dechlorinating the amidic nitrogens [254–256]. Chlorination-promoted hydrolysis is the second mechanism, based on the hydrolysis of the amidic group [198, 257]. Chlorine attack on the amidic N polarizes the amidic C and thus facilitates the hydrolysis of the C-N bond by a nucleophile, most likely OH^-. Under alkaline conditions, N-chlorination and dechlorination are believed to be the main processes occurring [252].

A part of the added chlorine reacts with organic and inorganic compounds present in the feed water. The reactions with ammonia and organic nitrogen are of utmost importance as they can affect the efficiency of the chlorination process. Chlorine reacts with ammonia in a stepwise reaction to form chloramines. The germicidal effect of chloramines is, however, much lower than that of chlorine and ozone, and PA-membranes may have a higher tolerance toward chloramines [194]. However, the main advantage of using chloramines as disinfection agents is the milder environmental impact of the DBPs.

Chlorine should be inactivated prior to the RO module, since the PA layer is highly chlorine-sensitive. This dechlorination step can be executed by dosing sodium metabisulfite (SMBS) and sodium bisulfite (SBS) into the feed stream. Activated carbon can be used as well [258]. Generally, PA-membranes only have a chlorine tolerance level of 200–1000 ppm · h [194, 259]. Manufacturers recommend a maximum concentration of 0.1 ppm chlorine or a maximum oxidation-reduction potential of 300–350 mV in the feedwater reaching the PA membrane [194]. After passing through the RO units, the water will be re-chlorinated and mineralized before being pumped into the distribution network.

Chlorine dioxide (ClO_2) has less degradation on PA membranes than Cl_2 gas. ClO_2 is also intrinsically a weaker oxidant than HOCl, HOBr and ozone. However, as it acts as a gas dissolved in seawater, it will penetrate through the membrane, degrade even the internal biofilm more efficiently, and provide biocidal protection downstream [194].

5.1.2.2   Chemical TPH/TOC/AOC Destruction

As mentioned in the first section, TPH/TOC/AOC are the key indicators of organic & bio-fouling. Oxidants are normally dosed into the feed water to oxidize a wide spectrum of organics, and thus to mitigate the organics and biofouling. Ozone, a potent germicide, can oxidize organic matter, and provide effective reduction of color, odor, UV254 absorbance (specific UV absorbance at 254 nm, an index of aromaticity), thereby improving the feedwater quality [260–270]. However, it's noteworthy that ozone cannot be used in the case where feedwater contains high bromide. Since the formation of bromate, a probable human carcinogen, is induced by reaction of ozone and bromide [200].

UV radiation is gaining more attention due to the concerns over hazardous DBPs introduced by chlorination and ozonation [201, 202], since it does not generate toxic residuals (as does ozone or chlorine). UV is normally combined with different radical promoters, H_2O_2, Cl_2, and O_3, etc., to oxidize the undesired organics, also known as AOPs (advanced oxidation processes) [249]. The combination of UV-irradiation and H_2O_2 leads to the photolytic cleavage of H_2O_2 into two ·OH. However, the molar absorption coefficient of H_2O_2 is relatively low. In the UV/O_3 process, UV irradiation (l < 300 nm) results in cleavage of dissolved ozone, followed by a fast reaction of atomic oxygen with water to form a thermally excited H_2O_2. Subsequently, the excited peroxide decomposes into two ·OH. Direct oxidation by a combination of ozonation and photolysis covers a wide range of organics. However, low energy efficiency of radical generation might explain that to the best of our knowledge, no published data on full-scale UV/O_3 applications are available [271]. UV/Cl_2 is another promising AOP, where UV-activated chlorine forms radical species, i.e., Cl· and ·Cl_2^- and ·OH which then oxidize the target compounds. Cl· is a more selective oxidant than ·OH, since it reacts favorably with electron-rich contaminants [272]. However, Cl· based reactions involve the formation of oxidative chlorine species (e.g., $ClO\cdot$, $OCl\cdot$), which might be oxidized by $\cdot OH$ to chlorate, perchlorate and halogenated OBPs [249]. A significant obstacle of UV based oxidation/ photolysis is the turbidity and TSS, as the high turbidity and TSS feed water will hinder the UV light transmission, and further inhibit the oxidation.

5.1.3 MINERAL AND METAL SCALE INHIBITION

5.1.3.1 Anti-Scalants

Scaling is triggered by the increased concentration of scale-forming species beyond their solubility limits and their precipitation onto the membranes. Normally, an increased concentration of scale-forming species in the bulk occurs due to permeate withdrawal, which is further enhanced in the region near the membrane surface by the superimposed effect of concentration polarization. This effect is more pronounced at high permeate fluxes and at low cross-flow velocities. In addition, scaling may also lead to physical damage of the membrane due to the difficulty of scale removal and to irreversible pore plugging. Therefore, mitigation of scaling is an important consideration in the operation of most RO and NF processes.

The most common constituents of scale in membrane applications are $CaCO_3$, $CaSO_4 \cdot 2H_2O$, and silica, while other potential scaling compounds are, e.g., $BaSO_4$, $SrSO_4$, CaF_2, calcium phosphates, ferric and aluminum hydroxides and ferrous sulfide [210, 273]. Barium and strontium precipitates tend to be less important because the cations are present in low concentrations, as compared to calcium [13]. Chemicals called antiscalants are used in brackish water RO systems to prevent such scaling. Advantages of using antiscalants are [274]:

- Preventing the need to replace expensive membranes prematurely

- Maintaining efficient plant operation at the highest possible rate of recovery

- Eliminating or reducing the use of hazardous acids

- Reducing water consumption by safely operating at high recovery rates

- Using fewer chemicals

- Producing less concentrate, and allowing for better environmental acceptance of that concentrate

- Reducing energy costs

- Reducing the downtime caused by frequent membrane cleanings

Scale inhibition effectiveness depends on the capability of an additive to interfere with the scale formation steps, i.e., either with the step of nucleation or with that of crystal growth. Various anti-scalants are considered to act according to one (or more) of the following main mechanisms of interference with crystal growth: (i) threshold effect (i.e., influencing the initial clustering process of the protonuclei), (ii) crystal distortion effect, (iii) dispersion, and (iv) chelation [210–212].

Calcium Carbonate

Calcium carbonate has three crystalline forms [275]: calcite (rhombohedric structure), aragonite (orthorhombic) and vaterite (hexagonal). However, vaterite, the third most common calcium carbonate polymorph, has seldom been detected in scale deposits, probably due to the presence of magnesium in the feed water [210]. The most common anti-scalants that suppress calcium carbonate are polymeric organic compounds, usually phosphonates or polyelectrolytes, such as polyphosphates, polyphosphonates and polycarboxylic acids [275]. Comparative performance of tetraphosphonates and diphosphonates toward calcium carbonate inhibition has been studied [276], which has elucidated that phosphonate additives prevent crystal growth by interacting with the calcium carbonate crystal surfaces. Besides, among the family of tetraphosphonates, the most efficient ones are those with the longer polymetylene bridging chains, except for the longest one—dodecamethylenediamine-tetrakis (methylenephosphonic acid). In terms of the diphosphonate family, the inhibitor efficiency increases with elongation of the $-CH_2$ chains. However, starting with the octamethylenediamine-tetrakis (methylenephosphonic acid) there is a reverse trend [276]. Trace amounts of zinc ions were also found effective in controlling $CaCO_3$ [211].

Calcium Sulfate

Another main role in RO inorganic scale is calcium sulfate. Calcium sulfate deposition from an aqueous solution appears in three different forms, where gypsum ($CaSO_4 \cdot 2H_2O$) is the major presence in RO fouling. Various anti-scalants have been studied and tested to inhibit gypsum formation. Dihydrate and hemihydrate forms of calcium sulfate have strong interactions with inhibitors like polyvinyl sulfonate (PVS) with respect to change of the crystal habit. The effective interaction among the negatively charged groups on the inhibitor and Ca^{2+} imparts changes in precipitate morphologies in the crystal lattice of calcium sulfate hemihydrates and dihydrate and scale inhibition [277]. Carboxylate anti-scalants are usually less effective in the presence of multi-valent metal ions like Ca^{2+} [278]. Higher temperature is found to help [279]. The effectiveness of common anti-scalants can be listed as following: formulated Polyelectrolyte > Polyacrylate > Hexametophosphate > > Pyrophosphate ≈ Tripolyphosphate ≈ Polystyrene Sulfonate ≈ Polyacrylamide ≈ Control (No Anti-scalant).

Phosphorus-based anti-scalants are quite popular in practice due to their remarkable performance; however, they may trigger environmental side effects. Polyphosphate additives discharged into the ocean act as nutrients for algae, leading to local flourishing and influence over the marine biota picture at intakes and outfalls [278]. Polyamino polyether methylene phosphonate (PAPEMP), an anti-scalant that inhibits calcium sulfate and calcium carbonate scales, help bacteria thrive in the phosphate environment [280]. Poly(acrylic acid) and poly(acrylamide), although common anti-scalants, due to their lack of biodegradability, also show negative impacts on the environment [278].

Eco-Friendly Anti-Scalants

Eco-friendly anti-scalant systems therefore should meet the following criteria: excellent scale inhibition, low aquatic and human toxicity, high biodegradability, good price/performance ratio, and free of phosphorus, nitrogen, and heavy metals [281]. Acrylic acid-allylpolyethoxy carboxylate (AA-APEC), a non-phosphorus anti-scalant, has a strong ability to inhibit calcium sulfate at a dosage of 2 mg/L, showing approximately 83.6% inhibition [282]. Poly(amino acids), with their amide-peptide linkage, are completely biodegradable and have considerable potential as green anti-scalants. Polysuccinimide (PSI) of various molecular weights has been synthesized from aspartic acid, with tested inhibition performance as follows: polyaspartate > poly(aspartamide-co-aspartate) > (SI-co-aspartate) > poly(SI-co-aspartamide) > polyaspartamide [278]. CO_2 has been found to be effective in scale inhibition including stopping calcite, aragonite and gypsum, compared to three commercial anti-scalants [283]. A group of researchers at the University of Crete in Greece have devoted efforts toward developing environmentally friendly anti-scalant additives [281, 284–287].

Silica

Silica has been awarded the "Gordian Knot" of water treatment due to the difficulty of either inhibiting or removing it. Silica (SiO_2) exists in crystalline and amorphous forms, between which crystalline silica has a very low solubility on the order of 6 mg/L SiO_2, whereas amorphous SiO_2 has a much higher solubility in the range of 100–140 mg/L SiO_2 [288]. Therefore, amorphous SiO_2 is the culprit in silica scale, which means those traditional scale control methods (inhibition and crystal modification) applied to crystalline mineral salt precipitates will be ineffective. The content of silicon in brackish water is commonly over 1000 ppm. Inevitably, such a high silicon level leads to undesirable silica fouling and significantly impairs the overall recovery [289, 290]. Silica scaling occurs through (a) condensation of monomeric silicic acid ($H_4Si(OH)_4$) on solid substrates containing the "–OH" group, (b) polymerized silicic acid or colloidal silica deposition, and (c) biogenic amorphous silica by living organisms [288, 289]. This reaction given in (a) is first-order and is catalyzed by OH^- in the pH range of 5–10, which tends to be boosted by the existence of metal ions, such as Fe^{2+}, Mg^{2+}, or Al^{2+} [289, 291]. Polymerization of silicic acids on the membrane surface, considered as the major mechanism of silica scaling, is mainly driven by the reaction between ionized and neutral silicic acid molecules [72, 292, 293]. This explains the observation that the rate of silica scaling is independent of both membrane hydrophilicity and free energy for heterogeneous nucleation. In contrast, membrane surface charge demonstrates a strong correlation with the extent of silica scaling [72].

Diverse anti-scalants have been applied to mitigate silica scale. The polyamide amine (PAMAM) dendrimers have proved to be effective [294, 295]. PAMAM dendrimers with –NH₂

Table 5.3: Silica anti-scalants [296, 297]

Anti-scalant	% SiO_2 Polymer Inhibition
Acusol TM 880 (HEUR)	90
Acrysol TM RM-845 (HEUR)	89
AquaZol TM 50 (poly(2-ethyl-2-oxazoline)	81
Triton TM CF-32 (EO/PO tert C2-alkylamine)	80
Dowfax TM DF-111 (alkyldiphenyloxide disulfonate)	77.2
Acumer TM 5000/Triton TM CF-32 (wt ratio: 17/83)	77
Dowfax TM DF-147 (alkyldiphenyloxide disulfonate)	75.9
Dowfax TM DF-122 (alkyldiphenyloxide disulfonate)	75
Dowfax TM DF-142 (alkyldiphenyloxide disulfonate)	74.5
Dowfax TM DF-101 (alkyldiphenyloxide disulfonate)	70.8
Tergitol TM L62 (nonylphenolethoxylate)	69.9
Control (no anti-scalant)	42

termini are more effective than their –COOH terminated analogs. However, NH_2-terminated PAMAMs also act as SiO_2 aggregators forming SiO_2–PAMAM composites. This inhibitor entrapment within the SiO_2 matrix induces loss of efficiency over time [294]. Polyethyloxazolines are also effective SiO_2 inhibitors, but their performance is less dependent on structural features [289]. Non-ionic polymers such as poly(2-ethyl-2-oxazoline), poly(vinyl pyrrolidone), ethylene oxide-propylene oxide block copolymer show mediocre to excellent performance as silica polymerization inhibitors [288]. Commercial silica anti-scalants have been tested as summarized in Table 5.3.

Major Suppliers & Products
There are plenty of anti-scalant suppliers and products available in the market (Figure 5.4). In Table 5.4, we summarize some major anti-scalants that are designed for different types of water. Generally, they target a certain type or multiple specific foulants, since it is hardly possible to inhibit all the foulants with one single anti-scalant.

5.1.3.2   Chemical Hardness and Alkalinity Removal
Lime, Lime-Soda, High-lime
Precipitation softening processes are applied to reduce feed water hardness, alkalinity, silica, and other constituents. The water is treated with lime or a combination of lime and soda ash (carbonate ion). These chemicals react with the hardness and natural alkalinity in the water

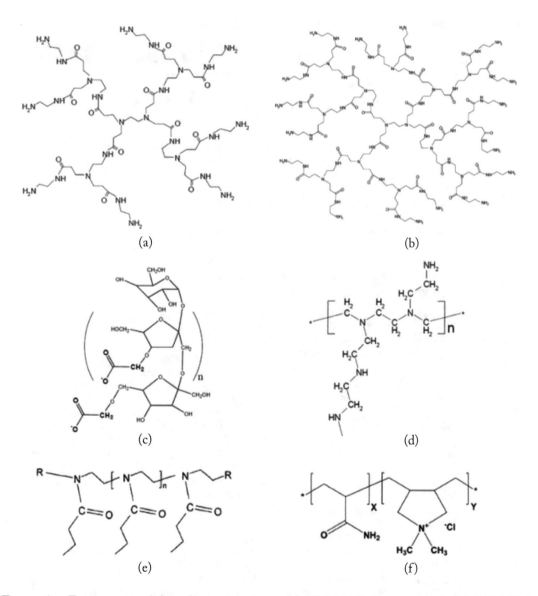

Figure 5.4: Environmental friendly anti-scalants: (a) PAMAM, Generation 1, (b) PAMAM, Generation 2, (c) carboxy-methylinulin biopolymer (CMI), (d) polyethyleneimine (PEI), (e) polyethyloxazoline (AQUAZOL), and (f) poly(acrylamide-codiallyldimethylammonium chloride) (PAMALAM) (reproduced from [184]).

Table 5.4: Major anti-scalant suppliers and products (*Continues.*)

Supplier	Product	Waters	Features
Avista	Vitec® 1100	Municipal wastewater	Inhibit calcium carbonate, metal oxides, silica, and mineral scale
	Vitec® 1200	Brackish waters	Inhibit iron metal oxides, silica, and mineral scale
	Vitec® 1400	Municipal wastewater	Inhibit calcium carbonate, calcium sulfate, metal oxides, silica, barium and strontium scale
	Vitec® 1500	Brackish waters	Inhibit calcium fluoride, calcium sulfate, calcium carbonate, calcium phosphate, barium sulfate, and strontium sulfate scale
	Vitec® 2000	Ground water/ Brackish waters	High level of iron oxides inhibition
	Vitec® 4000	Drinking water production	High level of silica inhibition
	Vitec® 5000	Drinking water production	Control scale and inorganic fouling in a variety of feedwater pH ranges
	Vitec® 7000	Brackish waters	High level of sulfate inhibition
	Vitec® 7400	Brackish waters	High level of silica inhibition
	Vitec® 8200 green	Drinking water production	Non-phosphorus
BWA	Flocon® 135	Brackish waters	Carbonate and sulfate scales
	Flocon® 190	Brackish waters	Compatible with polymeric coagulants
	Flocon® B38	Municipal feedwaters	Effective at low concentrations against a broad spectrum of microorganisms
	Flocon® Plus N	High salinity brackish water/ Seawater	Suitable for boron removal processes
	Flocon® 230	Brackish waters	Scale inhibition of carbonate, sulfate and fluoride scales and good dispersancy performance against iron, silica, organics, silt and clay
GE-Betz	Hypersperse MDC151	Brackish waters	Control scales including calcium phosphate, calcium carbonate up to LSI +3.0, calcium sulfate, barium sulfate, and strontium sulfate
	Hypersperse MDC220	Brackish waters	Control scales including calcium carbonate up to LSI +3.0, calcium sulfate, barium sulfate, and strontium sulfate
	Hypersperse MDC714	Brackish waters/ Drinking water production/Seawater	Control scale including calcium carbonate, calcium sulfate, and barium sulfate
	Hypersperse MDC775	Brackish waters	Control scales including silica, calcium carbonate, calcium sulfate, barium sulfate, and strontium sulfate
	Hypersperse MDC704	Brackish waters/ Seawater	Non-phosphorus; Control scales including silica, calcium carbonate, calcium sulfate, barium sulfate, and strontium sulfate
	Hypersperse MSI310	Brackish waters	High silica inhibition
AWC	AWC® A-101	Brackish waters	Inhibits calcium carbonate scale up to a Calcium Carbonate Nucleation Index (CCNI) of 2.2

Table 5.4: (*Continued.*) Major anti-scalant suppliers and products

Supplier	Product	Waters	Features
AWC (cont.)	AWC® A-102 PLUS/ULTRA	Ground water/ Brackish waters	Inhibits formation of calcium sulfate, calcium fluoride, barium sulfate, strontium sulfate and silica; Stabilizes metal ions to prevent metal oxides precipitation and disperses existing metal oxides/hydroxides, silt and clay particles
	AWC® A-103	Brackish waters	Especially designed for use where silica levels are high, up to 250% saturation of silica in the reject Effectively disperses aluminum-silicate foulants such as Kaolinites
	AWC® A-104	Brackish waters/ Drinking water production	Inhibits formation of all carbonate and phosphate scales; Phosphorus-free and environmentally compatible
	AWC® A-105	Ground water/ Brackish waters/ Drinking water production	Effectively stabilizes metal ions and disperses insoluble metal oxides in the reject
	AWC® A-106	Oil& gas production/Drinking water production	Inhibit inorganic scale formation and disperse natural organic matter (NOM) in membrane separation processes
	AWC® A-107	Municipal wastewater/Brackish waters/Drinking water production	Controls calcium carbonate scale formation and maintains good scale inhibition even in the presence of ferric coagulant carryover
	AWC® A-108Fe	Municipal wastewater/Ground water	Designed for use with RO/NF feed water that contains high levels of iron in the ferric state (oxidized iron), and aluminum
	AWC® A-109	Brackish waters/ Ground water	Helps reduce or eliminate acid dosing in RO/NF feedwaters that contain high alkalinity, even when operating at recoveries ≥ 90%
	AWC® A-110	Brackish waters	Effective silica inhibition
Genesys	Genesys LF	Brackish waters	Effective against a wide range of inorganic scaling species and also sequesters trivalent metal cations such as iron, manganese and aluminum
	Genesys CAS	Ground water/ Brackish waters	Inhibit high levels of Calcium sulphate
	Genesys SI	Ground water/ Brackish waters/ Drinking water production	High silica inhibition
	Genesys PHO	Waste water reuse/ agricultural leachate	Inhibit calcium phosphate and calcium carbonate scaling
	Genesys WB	Municipal wastewater/Brackish waters/Drinking water production	Contains dechlorinator to prevent membrane oxidation
	Genesys MP	High salinity brackish water/ Seawater	Inhibit inorganic scale, particularly calcium carbonate, iron and silica

Table 5.5: Typical softener effluent analyses [204]

	Raw Water	Removal of Calcium Alkalinity Cold-Lime	Lime-soda Softening (Cold)	Lime-soda Softening (Hot)	Lime Softening (Hot)
Total hardness (as CaCO$_3$), ppm	250	145	81	20	120
Calcium hardness (as CaCO$_3$), ppm	150	85	35	15	115
Magnesium hardness (as CaCO$_3$), ppm	100	60	46	5	5
Silica (as SiO$_2$), ppm	20	19	18	1–2	1–2
pH	7.5	10.3	10.6	10.5	10.4

to form insoluble compounds. The compounds precipitate and are removed from the water by sedimentation and, usually, filtration. Waters with moderate to high hardness and alkalinity concentrations (150–500 ppm as CaCO$_3$) are often treated in this fashion [204]. Lime softening can be used to remove carbonate hardness by adding hydrated lime:

$$Ca(HCO_3)_2 + Ca(OH)_2 \rightarrow 2CaCO_3 + 2H_2O \qquad (5.1)$$

$$Mg(HCO_3)_2 + 2Ca(OH)_2 \rightarrow Mg(OH)_2 + 2CaCO_3 + 2H_2O. \qquad (5.2)$$

Noncarbonate calcium hardness can be further reduced by adding sodium carbonate (soda ash):

$$CaCl_2 + Na_2CO_3 \rightarrow 2NaCl + CaCO_3. \qquad (5.3)$$

After the lime-soda process, the pH needs to be raised to keep silica soluble up to the target recovery concentration factor. Softening appears to only need about 3–4 hours HRT.

With lime softening, barium, strontium, and organic substances are also reduced significantly. The process requires a reactor with a high concentration of precipitated particles serving as crystallization nuclei. This is usually achieved by upflow solids-contact clarifiers. The effluent from this process requires media filtration and pH adjustment prior to the RO elements. Iron coagulants with or without polymeric flocculants (anionic and nonionic) may be used to improve the solid-liquid separation. Lime softening should be considered for brackish water plants larger than 200 m^3/h (880 gpm) [298]. Table 5.5 gives analyses of a typical raw water treated by various lime and lime-soda softening processes.

Compact Accelerated Precipitation Softening
Compact Accelerated Precipitation Softening (CAPS) is a hardness removal process in which feedwater pH is adjusted to 8–10.5 to reduce calcium and carbonate alkalinity by accelerated

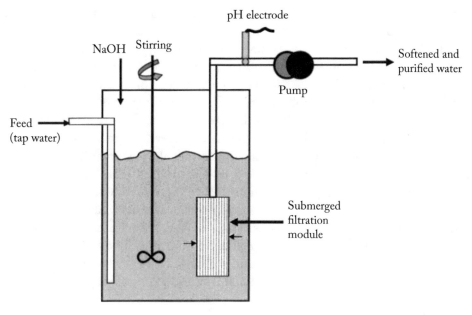

Figure 5.5: Schematic diagram of CAPS (reprinted with permission from [222]).

CaCO$_3$ nucleation and growth, which occurs (1) in the slurry due to crystal growth on the suspended particles and (2) in the filter cake due to (a) secondary nucleation and (b) crystal growth on the pore wall of the cake [218–223]. It consists of a mixing reaction tank, microfilter media, and a pump for recycling between the reaction tank and the filter media and for pumping the reaction suspension across the filter media, as shown in Figure 5.5.

Gilron et al. [220] compared CAPS efficacy for desalting brackish groundwater from the Negev region of Israel (original pH = 7.8) to dual media filtration (DMF) with acid or anti-scalant. As a result, CAPS removed 92–96% of the calcium at a flux of 40 L/h (equivalent of 2000 L/m^{2}h), achieving nearly a 90% recovery and reducing the requirement of membrane cleaning, whereas the media filtration only achieved a 72% recovery.

5.1.3.3 Silica Removal
Warm Lime Softening
Deposition of silica and silicate based foulants in BWRO systems is one of most difficult challenges for achieving high recovery. The complexity of controlling silica stems from the limited solubility of both amorphous (polymerized) silica and metal silicates in the feed waters. Once formed, silica scale is extremely difficult to remove and often requires the use of mechanical and/or chemical methods [299] including hydrofluoric acid-based chemicals which present environmental and safety concerns [300].

The most common method of silica removal from raw water is by precipitation with polyvalent metal hydroxides; $Fe(OH)_3$, $Al(OH)_3$, and $Mg(OH)_2$ are effective in removing both soluble and colloidal silica [299, 301]. The warm lime softening process operates in the temperature range of 49–60°C. It's noteworthy that in any warm lime or warm lime-soda ash process, temperature control is critical because temperature variations of as little as 2°C/h can cause gross carryover of the softener precipitates [204]. The silica reduction is accomplished through adsorption of the silica on the magnesium hydroxide precipitate. If there is insufficient magnesium present in the raw water to reduce silica to the desired level, magnesium compounds (such as magnesium oxide, magnesium sulfate, magnesium carbonate, or dolomitic lime) may be used. The optimal pH for silica adsorption onto $Mg(OH)_2$ is around 10–11, which coincides nicely with the conditions created during lime softening [301, 302]. Cold or warm process softening is not as effective as hot process softening for silica reduction. However, added magnesium oxide and good sludge contact will improve results. With the hot lime silicic acid removal process at 60–70°C, silica can be reduced to 1 mg/L by adding a mixture of lime and porous magnesium oxide [298]. Variations in raw water composition and flow rate also make control of this method difficult since it involves adjusting the amounts of lime and soda ash being fed [224].

Sodium Aluminate

Sodium aluminate is a combination of sodium oxide (Na_2O) and aluminum oxide (Al_2O_3, alumina). The formation of fresh alumina in the presence of dissolved silica leads to hydrolysis, formation of $Al_2O_3/Al(OH)_3$ precipitate and SiO_4 adsorption onto the precipitate. It is suggested to use 4 mol of Al per mole of dissolved silica to be removed, the optimum pH of which is between 8 and 9 [299, 301–304]. Besides, the dissociation of sodium aluminate in water generates OH^-, which promotes the formation of $Mg(OH)_2$ and further enhances the removal of silica. Sodium aluminate also reduces hardness, as described in Table 5.6.

Al-Rehaili et al. [305] evaluated the performance of lime-soda ash softening with coagulant aids (sodium aluminate or ferric chloride, and polymer) at six water treatment plants serving Riyadh, the capital of Saudi Arabia. For the low silica water source, the dose combination of lime (110 mg/L), soda ash (220 mg/L) and alum (30 mg/L) was adequate for silica control; while using caustic soda alone required a dosage of 125 mg/L for the same level of silica removal. For high silica waters, the dose combination of lime (90 mg/L), soda ash (300 mg/L), sodium aluminate (25 mg/L) and anionic polymer (0.05 mg/L) was adequate for silica control; a dosage of 175 mg/L caustic soda alone was adequate for the same level of silica removal. The removal of silica by adding caustic soda is attributed to formation of $Mg(OH)_2$.

According to Rajesh et al. [306], if silica is at 100 to 150 ppm and most of it is reactive silica, the solutions proposed are: (1) addition of sodium aluminate and (2) addition of dolomitic lime. Lime or caustic is needed as this precipitation takes place at a pH of 11.5. As a bonus, partially softened water will be obtained as well. The downside is the creation of a huge amount

Table 5.6: Lime-soda softening and silica reactions [301]

Process	Reaction	
Removal of CO_2	$CO_2 + Ca(OH)_2 \rightarrow CaCO_3\downarrow + H_2O$	(5.4)
Removal of carbonate hardness by lime	$Ca(HCO_3)_2 + Ca(OH)_2 \rightarrow 2CaCO_3\downarrow + 2H_2O$	(5.1)
	$Mg(HCO_3)_2 + 2Ca(OH)_2 \rightarrow$ $Mg(OH)_2 + 2CaCO_3\downarrow + 2H_2O$	(5.2)
	$MgCO_3 + Ca(OH)_2 \rightarrow CaCO_3\downarrow + Mg(OH)_2\downarrow$	(5.5)
Removal of calcium non-carbonate hardness	$CaCl_2 + Na_2CO_3 \rightarrow CaCO_3\downarrow + 2NaCl$	(5.3)
	$CaSO_4 + Na_2CO_3 \rightarrow CaCO_3\downarrow + Na_2SO_4$	(5.6)
Removal of magnesium non-carbonate hardness	$MgCl_2 + Ca(OH)_2 \rightarrow Mg(OH)_2\downarrow + CaCl_2$	(5.7)
	$CaSO_4 + Na_2CO_3 \rightarrow CaCO_3\downarrow + Na_2SO_4$	(5.8)
Sodium aluminate dissociation	$Al_2O_4{}^{2-} + 2H_2O \rightarrow Al(OH)_3\downarrow + 2OH^-$	(5.9)
Soluble silica with Mg hardness	$Mg(HCO_3)_2 + 2H_4SiO_4 \rightarrow$ $MgSi_3O_6(OH)_2 + 6H_2O + 2CO_2$	(5.10)
Hardness removal by sodium aluminate	$Na_2Al_2O_4 + Mg(HCO_3)_2 + 4H_2O \rightarrow$ $2NaHCO_3 + Mg(OH)_2\downarrow + 2Al(OH)_3\downarrow$	(5.11)
	$Na_2Al_2O_4 + Ca(HCO_3)_2 + 2H_2O \rightarrow$ $2Na_2CO_3 + CaCO_3\downarrow + 2Al(OH)_3\downarrow$	(5.12)
	$Na_2Al_2O_4 + 2CO_2 + 4H_2O \rightarrow 2NaHCO_3 + Al(OH)_3\downarrow$	(5.13)
Removal of silica as aluminosilicate	$Na^+ + Al^{3+} + 3H_4SiO_4 \rightarrow NaAlSiO_3O_8 + 4H_2O + 4H^+$	(5.14)

of lime sludge that needs to be dewatered and disposed of. Also, if the clarification or filtration system is not properly designed, it will lead to high SDI for downstream RO units [306].

5.1.3.4 Physical Hardness Removal

Ion Exchange

Ion Exchange (IEX) is another option for hardness removal. The IEX resin interacts with solutions containing magnesium and calcium ions, the hard ions preferentially migrate out of solution to the active sites on the resin, being replaced in solution by sodium ions [213, 214]. Industrial water treatment resins are classified into four basic categories:

1. Strong Acid Cation (SAC)

2. Weak Acid Cation (WAC)

3. Strong Base Anion (SBA)

4. Weak Base Anion (WBA)

Anion and cation exchange resins can be used in a single completely mixed reactor to remove DOM and hardness simultaneously.

Combined ion exchange treatment achieved > 55% total hardness removal and 70% DOC removal [307]. Smith et al. [308] applied self-regenerating anion exchange resins that selectively remove and replace sulfate by chloride as a pre-treatment of BWRO. Due to prevention of gypsum scaling, 80% recovery is obtained for a brackish water composition representative of groundwater in the San Joaquin Valley in California containing approximately 5200 mg/L of total dissolved solids or TDS. IEX is often used instead of chemical softening because it avoids sludge disposal; however, brine regeneration and disposal tend to also be very expensive unless the influent water quality is very good—free of colloidal particulates and oil and grease, which will coat and block IEX resins, as well as soluble hydrocarbons that degrade IEX resins rapidly.

Lime softening produces voluminous sludge that needs to be disposed of, whereas ion exchange processes generate concentrated brine or mineral acid as a waste regenerant stream. Residuals management will continue to be a major concern with these processes [309]. Also, sometimes polishing IEX softeners are used in the high pH scenarios. However, it has to be acidified ahead, and then the pH needs to be adjusted back up behind the IEX filters, which often requires degassing to purge excess CO_2 as well.

Nanofiltration

Admittedly, MF can remove suspended solids and lower the SDI. While in terms of UF, not only suspended solids and large bacteria are removed, but also (dissolved) macromolecules, colloids and small bacteria. However, they have no ability like NF, which can successfully remove the hard ions and further hold back the formation of scaling. Besides hardness, NF is also capable of removing pesticides and nitrate, which further enhances the produced feed quality [310]. The hardness removal performance of NF in former researches is identically remarkable; however, it may differ in detail due to the various water sources and NF membrane properties.

The total hardness of geothermal water is reduced with an NF-270 membrane (Dow Filmtec) by 84.7–85.5% at a temperature of 17°C and by 78.2–81.5% at 30°C. A high rejection ratio is achieved for sulfate ions (99.6–99.7% at 17°C and 99.1–99.6% at 30°C), and for metasilicic acid it amounts to 16.73–18.34% and 4.01–9.79%, respectively. Even better rejection ratios of undesirable ions can be achieved by dosing the feed with a certain anti-scalant [311]. Using commercial NF membranes, UTC 20, as a hardness removal approach, retentions higher than 90% were found for multivalent ions, whereas monovalent ions were retained for about 60–70%. A rejection of 94% is found for calcium, remaining constant at high recovery [312]. At a pressure of 22 bar, the removal by NF of Ca^{2+}, Mg^{2+}, SO_4^-, HCO_3^-, and is 89.6%, 94.0%, 97.8%, 76.6%, and 93.3%, respectively [191]. In addition, NF can separate monovalent ions including Cl^-, Na^+, and K^+, whereas with only 10–50% removal efficiency [313]. Mohsen et al. investigated NF to treat brackish water collected from the Zarqa basin in Jordan. The results suggest NF is efficient for hardness removal and gives a remarkable water recovery of up to 95% [314]. The NF–FO–BWRO system can increase the recovery rate of brackish water to over 90% using a multi-stage membrane treatment system [208].

Electrodialysis (ED)

Electrodialysis, like its derivative technologies, is a separation process that drives the movement of ions through ion exchange membranes using an electric potential [315]. An ED system consists of alternating cation exchange membranes (CEM) and anion exchange membranes (AEM) between an anode and a cathode. When a salt solution is fed between the membranes, an electric potential between the electrodes is established [141]. Cations are attracted to the cathode and are able to pass through the CEMs and then are retained by the AEMs. On the other hand, anions migrate toward the anode and can pass through the AEMs, but not through the CEMs. As a result, some compartments are depleted of ions, while the other compartments become more concentrated in ions.

If the feed solution contains negatively charged particles or large organic anions, they would likely deposit on the surfaces of the AEMs. By reversing the polarity of the system, these compounds would then move away from the AEM and back into the feed stream. This is called electrodialysis reversal (EDR) (Figure 5.6), and is usually employed periodically to prevent fouling or to restore membrane properties after fouling [316]. When the polarity is reversed, flow streams are also reversed. During EDR, the concentrate and the dilute compartments switch, so there is always some loss of ED product water.

Since ED is a process to remove electrically charged particles from a solution through ion exchange membranes, driven by the electrical potential difference, it has been widely used for desalination of natural waters, treatment of industrial effluents, salt production and hardness removal [317]. It is observed that increasing the voltage, flow rate and concentration boosts the hardness removal [215]. Under appropriate operating conditions, a significant hardness removal rate (approaching 100%) can be achieved [215, 317, 318].

Besides hardness ions, ED is also capable of selectively removing nitrate and fluoride [319–321]. ED is often performed at a constant voltage, causing the membranes to be under-utilized initially during the cycle when higher currents could be sustained. Shah et al. [318] proposed a model which suggests that the batch completion time is inversely proportional to the time-averaged ratio of the applied limiting current density. Therefore, the voltage-control increases the production rate by achieving higher ratios than is possible with constant-voltage desalination, which also decreases the total specific energy consumption [318]. Admittedly, ED exhibits a remarkable hardness control ability. However, the installation, operation, and energy consumption of ED are considerable, taking factors such as limiting current density, ion-exchange membrane fouling into account [316, 322–326].

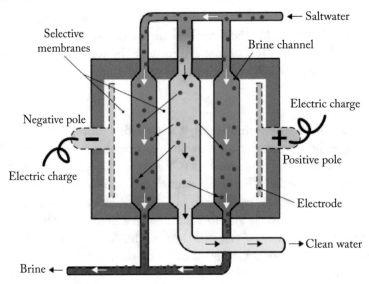

Figure 5.6: Schematic of electrodialysis. Anions migrate toward the anode and can travel through anion exchange membranes, but are retained by the cation exchange membranes. Cations experience the opposite effect. As a result, alternating compartments are depleted of ions, while the rest become concentrated in ions (reprinted from [327]).

Capacitive Deionization

Capacitive deionization (CDI) is an environmentally-friendly option for hardness removal which consumes a relatively small amount of energy and does not require any secondary regeneration wastes, expensive membranes, and high pressure pumps [217]. In a configuration of CDI cells, shown in Figure 5.7, the feed solution flows between each pair of high capacity electrodes, i.e., porous carbon electrodes. By polarizing both electrodes, the charged ions are adsorbed on each electrode surface; positively charged ions are attracted onto the surfaces of the negatively charged electrode and vice versa [217, 328, 329].

About 80% of Mg and Ca ions are removed at a high flow rate [217]. The performance is dependent on operational conditions and electrode materials [330–332]. Most of the electrodes are carbon-based materials, such as activated carbons, carbon aerogels, and carbon nanotubes [332]. With high specific surface area, these carbon-based materials will strongly adsorb a considerable amount of organic material, which may constitute nutrients supporting bacterial growth [331]. Consequently, due to bacterial growth, the CDI electrodes can become covered with a biofilm of bacterial extra-cellular polymeric substances (EPSs). This biofouling will impede electric charging of the electrodes, followed by loss of ion electro-sorption efficiency and an increase in energy consumption.

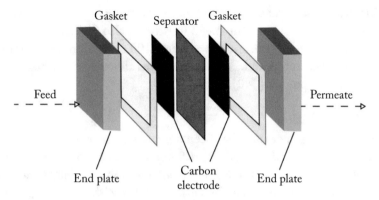

Figure 5.7: Schematic diagram of a CDI unit cell.

5.1.3.5   Metals (Fe/Mn) Removal

Green Sand

Oxidation and precipitation on pressure filters is the process used most commonly for the removal of metals, such as inorganic iron and manganese, typically found in ground waters with low dissolved oxygen. Pressure filters may use manganese greensand media, which has a high percentage of glauconite [2]. Passing brackish water, such as landfill leachate for instance [333, 334], through glauconitic greensand filters reduces the heavy metal cation content, lessens the unpleasant odor, and diminishes the murkiness of the leachate [335]. The capability of the greensand to trap metal cations is increased by prolonging the contact time between the leachate and the greensand. 95% metal removal efficiency can be achieved by greensand media under appropriate operating conditions [336].

Apart from greensand, other media materials such as zeolite, bone char and activated alumina [337–339] are also capable of metal removal. During the process, oxidants, such as chlorine, potassium permanganate, oxygen, or ozone are normally used to oxidize and precipitate the metal ions [1]. However, the application of chlorine in the metal removal process, which can disinfect as well, can trigger severe RO membrane degradation if there is any chlorine residues [194].

Metal-Catalyzed RO Membrane Degradation by Chlorine

Catalytic amounts of metals and transition metals, such as iron, zinc, aluminum, and copper, enhance membrane oxidation by increasing the oxidation potential of the oxidizing agents (e.g., chloramines and chlorine) [194, 196, 197, 199, 203]. It was claimed that the degradation mechanism of PA with and without metals is the same, although divalent ions catalyze membrane hydrolysis [340]. Aluminum was proven to have a more detrimental impact on membrane performance than iron, although their combination was even worse [25, 28]. Metal ions are believed to catalyze amide hydrolysis by the ability of metal-bound water to serve as a general acid catalyst in protonating the leaving nitrogen. The metal ions could also facilitate the breakdown of the C-N cleavage directly [341]. At concentrations below 100 mM, monovalent ions did not affect the membrane, while divalent ions did. At concentrations above this threshold, monovalent ions dominate the degradation process [340].

Metal-Catalyzed Silica Scaling

The polymerization of monomer silica has been accepted as the key factor to the formation of amorphous silica deposits [342]. However, direct deposition of mono-silicic acid or precipitation with multivalent metal (aluminum, iron) ions will result in membrane scaling as well [343]. Once deposited, silica scale is extremely difficult to remove without damaging the membrane. Dissolved silica can deposit as an impervious glass-like film on the membrane surface. In addition, iron and aluminum, even in trace amounts, may react with dissolved silica in concentrates to form metal-silicate scales. Iron and aluminum were found to be integrally incorporated into scale structures. Analysis of silica fouling deposits on sixty-six membranes showed the regular presence of metal ions, namely iron (88%) and aluminum (75%) in these deposits, while only one membrane fouled with silica had neither iron nor aluminum in the foulant [344, 345]. Sugita et al. [346] illustrated that the effect of aluminum on silica removal in geothermal brine is due to the formation of the aluminosilicate $(xAl(OH)_3 \cdot ySiO_2 \cdot nH_2O)$.

5.2   INTEGRATED MEMBRANE SYSTEMS

5.2.1   HERO/OPUS

A number of RO systems operating at high feed pH are effective for inhibiting silica scale formation. These high pH operation systems can be tailored to treat specific waters and to produce water with remarkable quality. Two commercial, patent-protected approaches to high pH operation are described below.

A high efficiency reverse osmosis (HERO) system [347] has gained considerable popularity due to its remarkable performance on silica removal [348, 349]. The HERO system is composed of reverse osmosis, a mixed bed ion exchange unit, MF, an ultraviolet sterilization unit, a decarbonation unit, and an electro-deionization unit [347]. In the HERO process, the feed water to the RO unit is raised to a pH value over 9, which suppresses the fouling of the RO membrane and has been found to achieve an extremely high recovery of about 99% [349]. The first step in a HERO process involves removal of Ca^{2+} and Mg^{2+} via weak acid cation exchange (WAC) media. The second step is decarbonation that reduces the buffering capacity of the water, thereby decreasing the amount of NaOH required to raise the pH. Next, NaOH is added to raise the pH to ≥ 9 before it enters the RO unit. During the RO treatment step, the pH of the concentrate increases to $\sim$11–11.2, which is the highest pH allowable for continuous duty of most RO membranes [21, 351]. The high pH operation converts dissolved silica into an anionic form (e.g., $H_3SiO_4^-$), which is repelled by the negatively charged membrane surfaces. In

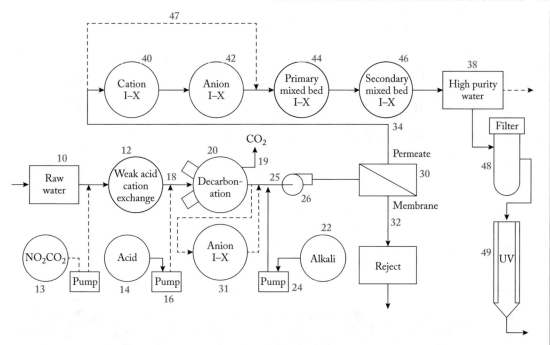

Figure 5.8: HERO™ technology apparatus (reprinted from [353]).

addition, the high pH on the concentrate side of the RO membrane prevents biofouling via both biocidal and electrostatic effects [11, 352]. HERO systems have been observed to operate with silica levels in the concentrate stream as high as 1500 mg/L without membrane fouling [348]. To produce water with extremely high purity at a high recovery rate for microelectronics and biopharma industries for instance, HERO could be a wise choice (Figure 5.8).

Another high pH operational system is the optimized pre-treatment and unique separation technology (OPUS®) [354], consisting of multiple treatment processes, involving chemical softening, media filtration, ion exchange softening and reverse osmosis (Figure 5.9). The pre-treatment processes ahead of the RO are designed to reduce the hardness, metals and suspended solids in the feed water. The RO process operates at an elevated pH, which effectively controls biological, organic and particulate fouling, eliminating scaling due to silica, and increasing the rejection of silica and boron. As shown in Table 5.7, the overall silica removal by the OPUS system is remarkably high, reaching 99.9% [354, 355].

OPUS II technology consists of multiple treatment processes involving chemical softening, membrane filtration, ion exchange softening and reverse osmosis operated at an elevated pH (Figure 5.10). The pre-treatment processes ahead of the RO are designed to reduce free oil, hardness, metals and suspended solids in the feed water. The RO process operates at an elevated pH, which effectively controls biological, organic and particulate fouling, eliminates

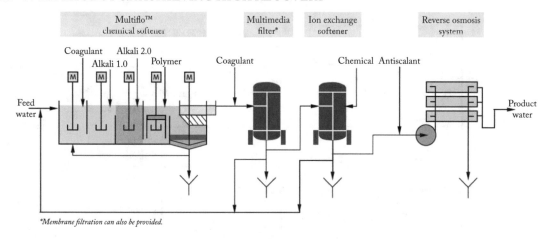

Figure 5.9: OPUS® technology apparatus (reprinted from [354]).

Table 5.7: OPUS® performance [352]

Constituent Type	Feed Water	First Pass Brackish RO Permeate	Second Pass Brackish RO Permeate	Percent Removal
Total dissolved solids, mg/l	6,500	538	25	99.6%
Total hardness, BDL as CaCO₃	242	<0.01	<0.5	>99.9%
Silica, mg/l	240	9.5	0.30	99.9%
Boron, mg/l	26	1.3	0.15	99.4%
Total organic carbon, mg/l	80	3.0	0.80	99.0%
Total suspended solids, mg/l	100	<1	<1	>99.9%

scaling due to silica and increases the rejection of silica, organics and boron. In the OPUS II technology, the feed water is subjected to chemical softening, free oil and solids removal in a pre-treatment process that uses CeraMem ultrafiltration technology. This process consists of a series of reaction tanks followed by a crystallization tank fitted with their patented Turbomix® mixing technology, which facilitates precipitation of hard ions and metals in the feed water and crystallization of the solids generated by precipitation. The softened water and crystalline solids are then processed through the ceramic membrane ultrafiltration system operated in crossflow mode for removal of free oil, bringing total hardness and suspended solids to lower concentra-

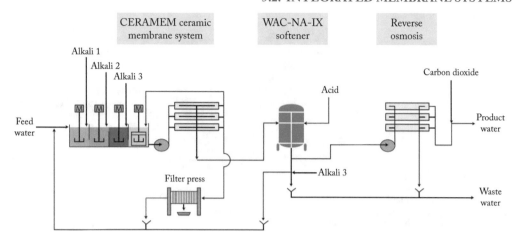

Figure 5.10: OPUS® II process (reprinted from [356]).

tions. The solid waste from the CeraMem process is continuously recycled to the crystallization tank and purged intermittently for dewatering and subsequent hauling to a landfill for disposal. The filtrate from the CeraMem process is further treated with ion exchange softening utilizing Weak Acid Cation (WAC) resin in sodium form for further removal of hardness and metals to lower concentrations, without the need for pH correction. The pretreated water is then pressurized through the RO, operated at an elevated pH in single or double pass mode, to reduce the TDS, boron and organics [356].

OPUS® has been successfully applied to the oil and gas production industry. It can be applied to de-oiling the water. In the case of the Chevron San Ardo project [357], the raw produced water is at a temperature of 200°F, and it contains about 25 ppm free oil, 80 ppm TOC, 240 ppm silica, 26 ppm boron, 240 ppm hardness and 6,500 ppm Total Dissolved Solids (TDS). The project goal was to reduce the feed water TDS to less than 510 ppm and boron to less than 0.64 ppm for recharge basins discharge, while achieving 75% water recovery across the treatment system and minimizing the volume of produced water requiring re-injection. The free oil removal system consists of an induced gas flotation system and a PowerClean walnut shell filtration system to reduce the free oil to less than 0.5 ppm. As shown in Figure 5.9, in the front, a MULTIFLO™ high-rate chemical softening process was applied to pretreat the feed water and raise the pH to 10–11. Then the water was pumped through the CeraMem® Ultrafiltration and passed with acid into the WAC-IEX resin. The well treated feed was then pumped into the double-pass RO system. The brine generated from the system was dewatered and disposed of in a landfill.

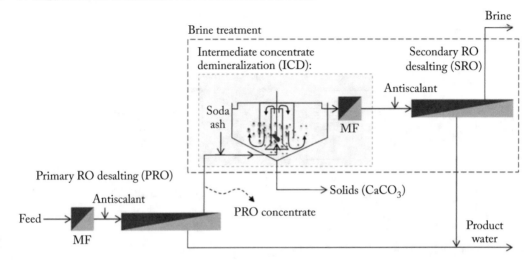

Figure 5.11: Scheme of inter-stage demineralization and secondary RO (reprinted with permission from [155]).

5.2.2 INTER-STAGE DEMINERALIZATION AND SECONDARY RO

Using intermediate concentrate demineralization and secondary RO desalting, with subsequent secondary RO desalting, is one potential approach to raise overall product water recovery [155], as depicted in Figure 5.11. In this process, an intermediate concentrate demineralization (ICD) step removes mineral scale precursors from the primary RO concentrate stream via chemical precipitation and subsequent microfiltration. The process involves continuous dosing of alkaline chemicals (e.g., caustic, soda-ash, and/or lime) for inducing $CaCO_3$ precipitation in a solids-contact reactor. As calcium and carbonate ions react to generate $CaCO_3$ solids, co-precipitation processes may occur, which may lead to the removal of other scale precursors such as barium, strontium, and silica [357]. After solids removal via sedimentation and microfiltration, the demineralized PRO concentrate is desalted in a secondary RO (SRO) step to recover product water and thus reduce the volume of the residual brine that must be disposed of. This brine treatment process can enhance the overall water recovery level of BWRO to over 95% [359–361].

5.2.3 MAXH₂O DESALTER

The MaxH$_2$O Desalter technology system, owned by IDE Technology, enables operation of a reverse osmosis (RO) system with feed water quality varying from 1,000–70,000 mg/L TDS, and at water recoveries from 25–99.9%. The system contains an RO system with an integrated salt precipitation and removal unit, similar to the inter-stage demineralization. The process provides a solution for difficult industrial wastewater and brackish water requiring high recovery

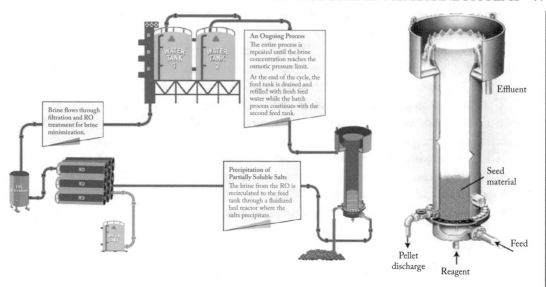

Figure 5.12: The brine minimization technology process scheme (reproduced from [362, 363]).

operation and successfully overcomes the different challenges, e.g., scaling of sparingly soluble salts, and organic and bio-fouling.

For the integrated salt precipitating unit, a pellet reactor is used. The pellet reactor removes only the salts that can harm the desalination process and, without the addition of chemicals, produces pellets of more than 90% dry solids content, which do not require further sludge dewatering treatment. In addition, the reverse osmosis unit is operated at high shear velocity, minimizing the negative effect of concentration polarization, thus improving the RO process efficiency. The process minimizes the brine quantity to the maximum threshold limit of the osmotic pressure, with minimum operational expenditures.

The operating principle of the technology is the recirculation of treated water through the RO system at high shear velocity, and continuous precipitation and removal of supersaturated salts from the recirculated brine. This significantly reduces the salt concentration build up near the RO membrane walls, thus preventing the precipitation of sparingly soluble salts on the membranes. The brine flows through the salt precipitation unit downstream to the RO membranes, where it reduces the saturation of sparingly soluble salts, thus enabling continuous cycling through the RO system, until reaching maximal brine osmotic pressure [362] (Figure 5.12).

In a BWRO brine treatment pilot project [363], a brackish water source with 8,000 micro S/cm, Silica 15–18 ppm, Calcium 184 ppm, Alkalinity 290 ppm as $CaCO_3$, with the presence of manganese, aluminum and iron caused significant problems for the plant. The original plant needs to have CIP and cartridge filters replacement every month. By applying MaxH$_2$O

Desalter, the total recovery of existing plant is enhanced from 72–95%, with the successful operation under the calcite scaling challenge.

In another case [363], the Jafora BWRO plant operated at 90% recovery under the challenges of silica and calcite scaling. The brine contained silica 230 ppm, calcium 1,350 ppm, alkalinity 2,290 ppm as $CaCO_3$. With a MaxH$_2$O Desalter, the recovery was pushed up to 97%. Calcium carbonate scaling was controlled in a very precise way, where the saturation index of calcium carbonate was kept below the maximum ability of the antiscalant. Optimization of the antiscalant dosing and type were required for the control of the silica scaling.

In the case where this system was used for treating mining water at a copper mine in Chile [363], 92% recovery was achieved under the threat of gypsum scaling. The source water contained calcium 301 ppm and sulfate 2020 ppm. Fifty percent of the sulfate was removed by a pellet reactor in the form of calcium sulfate. The removal process was induced by dosing with sodium carbonate.

5.3   NOVEL RO PROCESS MODIFICATIONS

5.3.1   FEED FLOW REVERSAL

Emerging modified processes can enhance the performance of RO systems and help achieve superior recovery. An innovative desalination technology called flow reversal, developed by Rotec LTD, has been proposed to be an alternative that averts membrane mineral scaling. The flow direction of the saline stream in RO pressure vessel arrays is periodically switched in a flow reversal process, as shown in Figure 5.13. By periodically switching the flow direction, the scale does not have time to form on the membranes surfaces before being swept away by under-saturated feed solution conditions. The frequency of switching is dictated by the time it takes for a supersaturated solution in the concentrate to grow a population of scale particles that can allow continued scale growth (i.e., induction time). By using the effectively under-saturated feed to sweep away the beginning scale particles in the concentrate before they exceed a critical size, extensive precipitation is prevented [362–364]. The flow reversal process increases recovery by 10–15%, and dramatically reduces brine volumes required for disposal by up to 70%.

Combined with the design of the block rotation, Figure 5.14, switching the second stage via rotating the blocks, the flow reversal technology is able to effectively alleviate the scaling and fouling on membranes. Therefore, a high recovery at low risk damaging the membrane is achieved. The downside includes the high complicity of such elaborated systems requiring well-trained operators to assure a decent running of the entire system. Also, the downtime of the system can be long due to the flow reversal operation and the block rotation, which, to a certain extent, affects the overall efficiency.

In the PUB project, Singapore, the AdEdge Water Technologies retrofitted an existing conventional RO system to the flow reversal system, the overall recovery was enhanced from 75–90%. The local price for the permeate was $1.22 USD/m^3, and the brine disposal cost $0.10 USD/m^3. Originally the permeate produced was 740 MGY (million gallons per year),

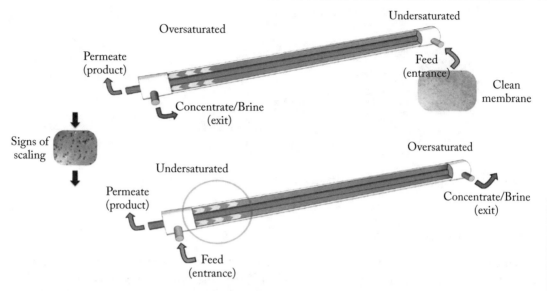

Figure 5.13: Flow reversal principle—switches the connection of feed and concentrate before supersaturated solutions can precipitate from the concentrate onto the membrane (reprinted from [367]).

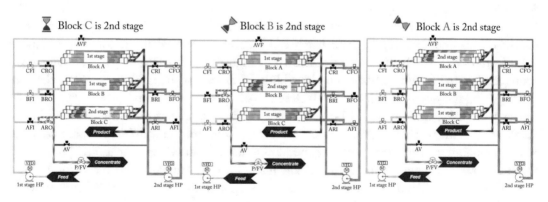

Figure 5.14: Block rotation (reprinted from [367]).

whereas the brine generated was 243 MGY. With the retrofitting flow reversal system, the permeate was increased to 872 MGY, and the brine was decreased to 98 MGY. Accordingly, the total savings of retrofitting the conventional RO to the high recovery RO were $665,000 per year [367].

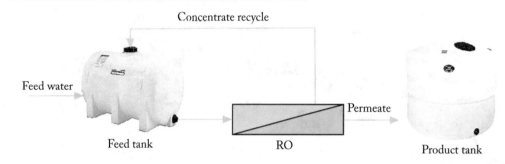

Figure 5.15: Simplified batch flow diagram.

5.3.2   BATCH, SEMI-BATCH

Batch and semi-batch processes are considered in theory as approaches that can obtain the thermodynamic minimum energy of desalination [366]. The batch RO process, shown in Figure 5.15, recycles the brine to an unpressurized feed tank until the target recovery or tank concentration is reached [369, 370]. This is in wide use in laboratory settings where it has demonstrated the potential to reduce the volume of concentrate produced during RO treatment with high concentrations of sparingly soluble salts without indications of membrane fouling or scale deposition [371].

A true batch RO system has yet to be demonstrated in which the system remains pressurized during operation [368]. Semi-batch RO is another practical alternative that maintains some features of a batch RO process as shown in Figure 5.16. In semi-batch RO, pressure is maintained in the feed/membrane loop by adding fresh feed at the same flow rate as the permeate exiting the system [368]. For BWRO, semi-batch processes can increase thermodynamic efficiency by 25%, while batch RO can double efficiency compared to continuous RO [372]. Single stage semi-batch RO processes can be designed to operate at 97% or higher recovery while maintaining compliance with the membrane manufacturer's specifications. Recovery is adjustable over a wide range and automatically maintained by standard system controls [373].

Desalitech's Closed Circuit Reverse Osmosis process [374, 375], a semi-batch process, is an elegant way of operating crossflow reverse osmosis membranes in a highly efficient and flexible simple filtration device. Like any simple filtration device, the Closed Circuit Reverse Osmosis system features equal feed and permeate flow rates during normal operation mode as shown in Figure 5.17. At a software based set point, the system automatically flushes out all the concentrate, and then returns to its normal operation mode. The flush is triggered by the Closed Circuit Reverse Osmosis operating software, based on any combination of flow, concentration, pressure and additional set points. During the concentrate flush step, the system continues to be fed and to generate permeate, while concentrate is pushed out of the system in one sweep. Crossflow for standard reverse osmosis membranes is attained with a single stage of

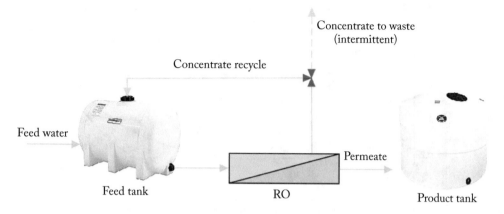

Figure 5.16: Simplified semi-batch or "feed and bleed" flow diagram.

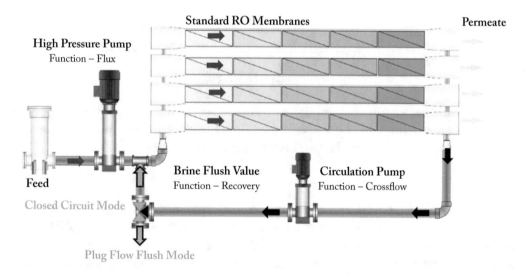

Figure 5.17: Closed circuit reverse osmosis (CCRO) (reprinted from [376]).

parallel membrane housings and a low pressure circulation pump that generates optimal cross flow conditions. Concentrate is recirculated to the membrane feed and recovery increases with each concentration cycle. Recovery is achieved in time with recirculation and not in space with multiple membrane stages in series. The overall recovery rate in the CCRO process is a function of the time between brine flushes. Therefore, it is not necessary to use multiple stages as is required to increase recovery in traditional RO processes. A high-recovery design is constructed in one stage typically consisting of four or five elements per vessel to optimally balance performance and costs [375].

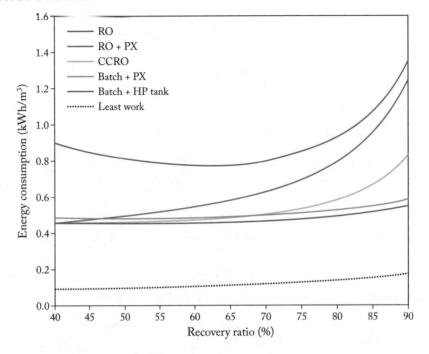

Figure 5.18: Modeled energy consumption of continuous and time-variant RO configurations for various recovery ratios with 3 g/kg NaCl feed. Least work of separation is also shown. PX stands for pressure exchanger and HP for high pressure (reprinted from [377]).

Although the least work of separation increases monotonically with recovery ratio as shown in Figure 5.18, the actual energy consumption of continuous RO without an ERD reaches a minimum around 60% recovery as a result of throttling high pressure brine as it leaves the system. At low recovery ratios, a larger amount of fluid is irreversibly depressurized per unit permeate than at high recovery ratios. While energy recovery devices can replace throttles and recover much of the loss, the time-variant systems reduce energy consumption by only pumping the permeate volume to high pressure, thus eliminating the need to (irreversibly) recover energy from the brine [377].

Figure 5.19 shows that continuous RO, CCRO, and batch RO have higher efficiency at higher feed salinity because the least work of separation rises, while the losses stay relatively fixed; however, the effect of recovery varies between them. Both CCRO and batch RO are energetically superior choices to continuous RO at low recovery. Batch RO is the most energy efficient of the three systems at high recovery. At most lower recovery ratios, CCRO and batch systems perform similarly. Only at the lowest salinities and recovery ratios, rarely a desirable combination, is continuous RO the most efficient choice [377].

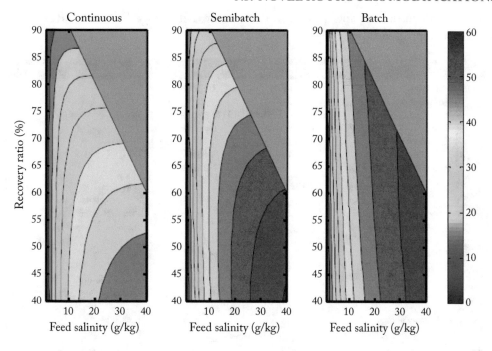

Figure 5.19: Exergetic efficiency (in %) for continuous (with pressure recovery), semi-batch (CCRO), and batch RO systems.

Coca-Cola has already installed a half-dozen of the CCRO systems, with more units currently under construction. These systems around the globe operate at recovery rates ranging from 91–95%, depending on the quality of their feedwater source and the specific bottling line. This greatly improves the 75–85% recovery rates that they were accustomed to with traditional multistage RO systems [375]. Another case study of CCRO is the Sonoran Desert paper production. The plant upgraded its original multistage RO to the CCRO, enhancing the recovery from 73–88%. Also, the permeate production doubled. CIP was extended from 24–4X/Year, as shown in Table 5.8. In a pilot study conducted in 2018, CCRO showed a 100% removal on PFAS, achieving 90% recovery [378].

5.3.3   IDE PULSE FLOW RO

In IDE's patented Pulse Flow RO (PFRO) [379] technology the flow is not stable, and the osmotic and feed pressure change frequently and rapidly (Figures 5.20 and 5.21). This method has six times the ability of the standard RO process to remove concentrated solute ions or fouling particles [379]. In between brine discharge pulses, the RO membrane operates in dead-end mode, with 100% recovery, which provides significant energy savings by eliminating the pressure

Table 5.8: Sonoran desert paper production case study: Multistage RO vs. CCRO, long-term performance (reprinted from [375])

		Multi-stage	Closed Circuit		Value
Reverse Osmosis Design	# of trains	1 × 100%	2 × 100%		
	# of stages	2	1		
	Array (per train)	3:1 (28 membranes)	10 (40 membranes)		
	Process	Steady-state	Dynamic		
	Recovery	73%	88%		
	Utilization rate	90%	63%		
	Permeate flux (gfd)	15.4	15.5		
System Performance	Daily process water (gal)	155,520	311,040	200%	Increase in permeate production
	Daily wastewater (gal)	57,521	42,414	26%	Reduction in wastewater generated
	Specific power consumption (kWh/kgal)	1.75	1.67	5%	Reduction in energy required
	Antiscalent consumption (ppm)	8	3	63%	Reduction in antiscalent use
	Biocide frequency	3X/week	1X/week	300%	Extension in biocide frequency
	CIP frequency	24X/year	4X/year	300%	Extension in CIP frequency
Membrane Performance	Lead element flux (gfd, avg)	20.5	18.0	12%	Reduction in lead element flux
	Flux distribution (gfd, avg)	(6.5–20.5)	(13.9–18.0)		
	Max beta value	1.14	1.09	4%	Reduction in max beta value
	Beta range	1.03–1.14	1.04–1.09		

drop per stage [379]. The pressure drop in each standard RO stage is usually approximately 1 to 2 bar, which results in a total pressure drop per RO train of 4–5 bar.

During the production cycle, the gauge and osmotic pressure interchange in a non-exact parallel pattern. Bacteria are exposed to rapid changes in gauge and osmotic pressure. The capacity of organisms to respond to fluctuations in gauge and osmotic environments is limited and such fast changes do not enable them to reproduce. However, the substantial cost of capital, the CIP frequency and module maintenance should be taken into consideration. Besides, the membranes applied in such processes should possess high tenacity under the high-pressure operation. In other words, the system is required to be well monitored.

Some advanced RO processes have been discussed in this section. Table 5.9 briefly summarizes the advantages and disadvantages of such designs. They are all purposely designed for high recovery at the cost of either high CAPEX or OPEX. Processes like CCRO, Flow Reversal and Pulse Flow try to beat the thermodynamics with kinetics, which means they try to inhibit or repress scaling or fouling, in their own ways, by elaborately dumping the highly saturated brine before it scales (after saturation scale formation requires an induction time) on membrane

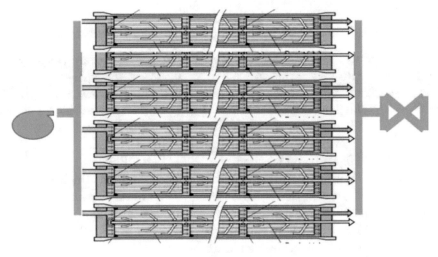

Figure 5.20: Pulse flow RO (PFRO) (reprinted from [379]).

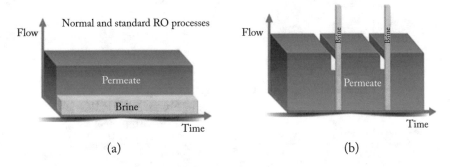

Figure 5.21: (a) Conventional RO: The flow is continuous and uninterrupted. (b) Pulse Flow RO: Brine flow discharges as pulses in a very short time at the maximum flow allowed by the membrane manufacturer (reprinted from [381]).

surfaces or periodically shifting the flow direction to sweep away the beginning scale particles. Considering the incredibly high cost of brine disposal, these innovations are expected to lower the overall cost in long-term operation and overcome the high scaling and fouling propensity featured by conventional multistage configurations.

Table 5.9: Advanced RO processes/patents and their advantages, disadvantages

Advanced Techniques	Advantages	Disadvantages	Ref.
HERO/OPUS	High recovery Combination of well-developed and established technologies; Higher boron rejection; Prevention of saturation of gypsum, calcite, barite, and silica fouling; Low RO scaling propensity; Capable of achieving remarkably high recoveries (up to 99%)	pH-shift not suitable for highly buffered waters; High CAPEX; High OPEX; Sodium Pump (higher permeability whereas lower rejection)	[348,381]
Flow Reversal	High recovery; Low scaling/fouling propensity; Remote control;	High CAPEX; Long downtime; High system complexity;	[365]
Pulse-flow RO	High recovery; Low biofouling; Low scaling/fouling propensity;	High OPEX; Mature RO monitoring required; Frequent CIP, RO maintenance;	[379]
Batch, Semi-Batch (CCRO)	High recovery; Low pressure; Extended CIP; Higher thermodynamic efficiency than continuous RO;	Limited operating pressure; Elaborated control system required; High CAPEX	[382]
MaxH2O Desalter	The low and independent instantaneous RO recovery, per cycle, that allows operation with a single stage RO system; Low volume brine; Low fouling propensity	High CAPEX; High OPEX; Solid disposal; Seeding reactor reactivation	[380]
Inter-stage demineralization and secondary RO	High recovery; Easy operation	High OPEX; Brine/Solid disposal High membrane cleaning frequency	[357–359]

CHAPTER 6

Methods for Concentrating and Treating Brine

6.1 EVAPORATIVE BRINE CONCENTRATION (EBC) FOR MINIMAL LIQUID DISCHARGE

6.1.1 THERMAL DISTILLATION

Typical evaporative processes, shown in Figures 6.1, 6.2, and 6.3, include mechanical vapor compression (MVC), multiple effect distillation (MED), multi-stage flash distillation (MSF), solar ponds and wind aided intensified evaporation (WAIV). In mechanical vapor compression (MVC), vapor generated during evaporation is recompressed using a mechanically driven compressor, which also acts as a heat pump, adding energy to the vapor. After compression of the vapor and condensation of the heating system, the condensate leaves the cycle with the heating steam separated from the vapor by the heat exchange surface of the evaporator [384]. The limitation of this type of technology is the maximum size of the volumetric compressors. Its maximum capacity does not allow for high production of desalinated water. In multi-effect distillation (MED), the evaporator consists of multiple cells operated at decreasing levels of pressure. The steam generated in one stage or effect is used to heat the concentrate in the next stage as the next stage is at a lower temperature or pressure. Similar to the multi-stage flash (MSF) process, the performance of the MED process depends on the number of stages or effects. MED plants normally use an external steam supply at a temperature greater than 100°C.

6.1.2 EVAPORATION IN SOLAR POND

Disposal in evaporation ponds is another conventional method of concentrate management for desalination facilities. It involves pumping the brine into shallow, lined basins and allowing water in the brine to evaporate naturally by solar irradiation. When water evaporates, salts from the brine accumulate in the pond and are removed periodically for disposal in landfills. Often, a series of ponds will be constructed to allow for continuous brine discharge. A key advantage of this method is that it is very simple: it does not require any advanced technology, and little maintenance is required for operation [385]. It is also a viable and proven disposal option for facilities in arid inland areas far removed from surface waters or WWTPs [385]. However, the use of evaporation ponds is very site-specific (Figure 6.4).

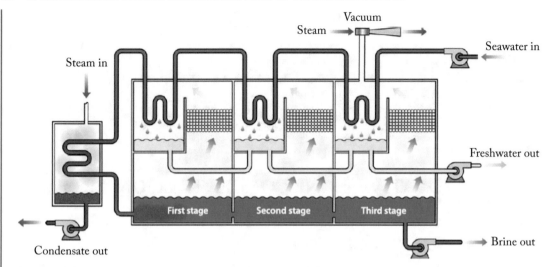

Figure 6.1: Schematic diagram of MSF unit (reprinted from [327]).

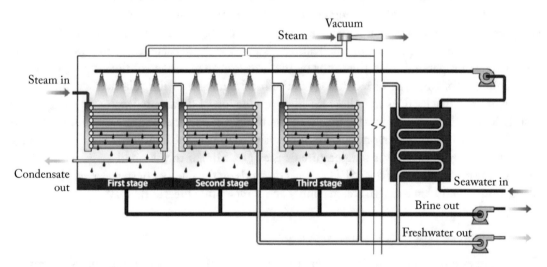

Figure 6.2: Schematic diagram of MED unit (reprinted from [327]).

While disposal in evaporation ponds is a simple concept, it is a very land intensive practice and requires flat terrain. The area of land required mainly depends on evaporation rates in the region and the flow rate of brine. Because this method depends on the natural evaporation of water, evaporation ponds are especially suitable for warm, dry climates that result in high evaporation rates. When evaporation rates are high, the required pond surface area decreases.

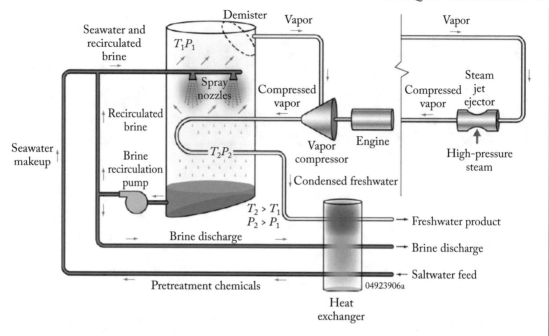

Figure 6.3: Schematic diagram of MVC process (reprinted from [327]).

In addition, a lower brine flow rate will require less pond area. Therefore, this method is most feasible for smaller RO plants in warm, arid climates where land costs are low.

A primary environmental concern with this method is that the high salinity concentrate could seep underground and contaminate groundwater sources. Therefore, regulations typically require an impervious lining for the basin, and monitoring wells to detect leaks. An appropriate lining material and thickness must be chosen so that no, or minimal, deterioration occurs in the high salinity waters. The impermeable lining is one of the largest expenses associated with evaporation ponds. The cost of land acquisition and adequate lining can drive the capital cost of evaporation ponds very high. This method has few economies of scale and is most economically feasible for low brine flows. Brine disposal to evaporation ponds is well suited for warm and dry climates where land is inexpensive and abundant, however site-specific.

6.1.3   WIND AIDED INTENSIFIED EVAPORATION (WAIV)

Wind aided intensified evaporation (WAIV) technology was developed as an alternative to evaporation ponds. Like evaporation ponds, WAIV involves the evaporation of water from brine to minimize the disposal volume. In WAIV, brine is continuously recirculated as falling films down vertical hydrophilic surfaces and exposed to wind [386]. Evaporation is driven by the vapor pressure gradient between the dry wind and the wet surfaces. Since vertical surfaces are

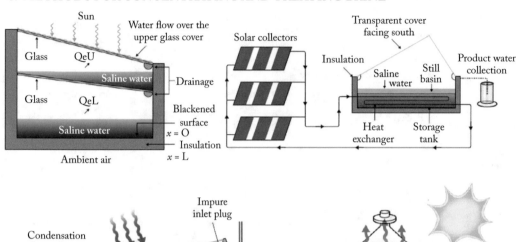

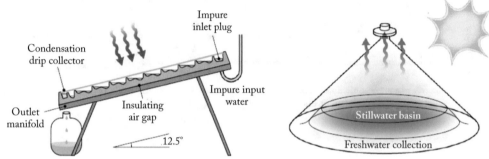

Figure 6.4: Several types of solar evaporation (reprinted from [327]).

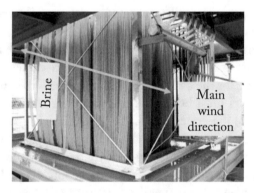

Figure 6.5: Image of a WAIV unit. Brine flows down the vertical surfaces and wind blows parallel to these surfaces, as indicated by the yellow arrows (reprinted with permission from [387]).

used, they can be closely packed together to drastically increase the evaporative capacity per footprint area [387]. The packing densities of the surfaces are typically 20 m²/m footprint and greater [388]. A WAIV unit can be seen in Figure 6.5. In some cases, WAIV can even reduce the required footprint area by an order of magnitude, compared to evaporation ponds.

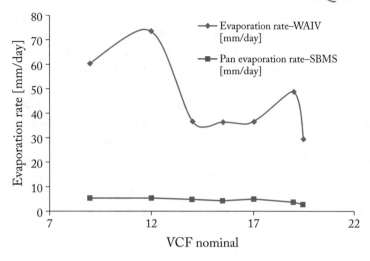

Figure 6.6: Comparison of evaporation rates of the WAIV unit compared to a conventional evaporation pond against the volumetric concentration factor (VCF) (reprinted with permission from [387]).

Katzir et al. conducted a pilot study to test the ability of WAIV to evaporate brackish water RO brine with 1.2% TDS, or 12,000 mg/L TDS [387]. As a result, the brine was concentrated to 10–15% TDS (100,000–150,000 mg/L TDS) and the evaporation rates of the WAIV unit were observed to be 10 times that of conventional evaporation ponds, which would allow for ponds that are 10 times smaller [387]. A comparison of the evaporation rates of the WAIV unit and an evaporation pond is show in Figure 6.6.

Compared to traditional evaporation ponds, WAIV technology significantly reduces the land requirements for brine disposal. In addition, energy consumption and operating costs are low in WAIV because evaporation is driven simply by wind dryness. These factors make WAIV suitable for dry, semi-arid regions where energy costs are high. On the other hand, WAIV also has the potential to pollute groundwater, and is not currently feasible for large volumes of brine. As of now, only bench and pilot scale studies of WAIV have been performed. Industrial scale experiments are necessary to study the expected decrease in efficiency with additional close packed surfaces [387].

6.1.4   PHYTODESALINATION

In phytodesalination, brine is beneficially reused to irrigate salt tolerant vegetation, usually halophytes. In turn, the halophytes can remove some of the salt from the saline water. While many plants do not grow well in the presence of salt, studies have shown that halophytes typically require some salt to achieve their maximum growth rate. Many halophytes have low to moderate growth rates and water consumption since they are usually found in high-stress environments. However, researchers have found that there are halophytes that can have high growth rates and water consumption, even with feed waters beyond 40 g/L TDS. They can be grown and used for landscaping, grazing, windbreaks, and development of wildlife habitats. Some can produce oilseed, grain, or fodder.

Soliz et al. studied the growth and nutritional value of *Atriplex lentiformis*, or quailbush, irrigated with RO brine in a desert irrigation district in the Sonoran Desert [389]. In the pilot study, the plants were grown in drainage lysimeter basins and irrigated with 2.6–3.2 g/L RO brine. The annual biomass yields were 15–22 tons/ha, similar to high-yield crops like alfalfa (20.3 t/ha) and Sudan grass (3.6 t/ha). The drainage fraction ranged from no drainage to 12% of the applied water. As livestock feed, the main value of *A. lentiformis* is probably as a mineral and protein supplement and as a forage replacement for ruminants, but feeding trials are necessary to verify this. From their research, Soliz et al. concluded that halophyte irrigation is an attractive beneficial use of brine at least at locations like arid zone irrigation districts [389]. However, as they suggest, the practicality and sustainability of this method is ultimately site specific.

Many improvements to the water treatment performance in this brine management method are still necessary. One study found that for a brine production of 16,500 m³/day, land requirements for phytodesalination would be over 200 hm², suggesting a need for higher land use efficiency [315]. In addition, brine irrigation could impair both soil productivity and groundwater quality. Even low TDS concentration brine can deposit large amounts of salt in the soil, which could reduce soil permeability [388]. Irrigation with high salinity brine can also increase the salinity of underlying aquifers. It can especially pose a threat to water quality if the brine contains arsenic, nitrates, or other harmful contaminants. If this type of method is allowed, it would typically require extensive monitoring to ensure that subsurface waters are protected and that salt does not accumulate in the soil [388]. Moreover, to meet regulatory quality standards the brine might need to be diluted before being applied to the land, which would not likely be possible in inland communities with no easily accessible fresh water sources.

6.2   MEMBRANE-BRINE CONCENTRATION (MBC) FOR MINIMAL LIQUID DISCHARGE

6.2.1   OSMOSIS-ASSISTED RO, OARO

By avoiding a phase change, non-evaporative membrane-based treatment technologies may reduce the energy intensity of brine dewatering processes. Reverse osmosis (RO), forward osmosis (FO), assisted forward osmosis or pressure assisted forward osmosis (PAFO) offer several pathways for brine dewatering across a semi-permeable membrane [17, 390–393]. A positive hydraulic pressure difference ($P_f - P_p$, ΔP) drives water transport, and a negative ΔP retards water transport. In contrast, a positive osmotic pressure difference ($\pi_f - \pi_p$, $\Delta \pi$) retards water transport, and a negative $\Delta \pi$ drives water transport.

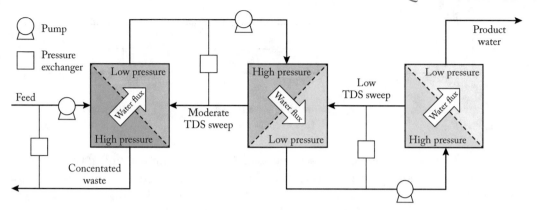

Figure 6.7: Schematic of OARO process. The first two modules are OARO modules, while the final stage represents an RO module (reprinted from with permission [393]).

In RO, a positive hydraulic difference $(+\Delta P)$ drives water transport against the retarding force of a positive osmotic pressure difference $(+\Delta \pi)$. In FO, there is little hydraulic pressure gradient $(\Delta P \approx (0))$ and a highly concentrated draw solution establishes a negative osmotic pressure difference $(-\Delta \pi)$ to drive water transport from the feed to the draw. When a positive hydraulic pressure gradient is used to augment this negative osmotic gradient driving force $(+\Delta P, -\Delta \pi)$, the process is labeled pressure assisted forward osmosis (PAFO). While not a separation process, pressure retarded osmosis (PRO) processes utilize the hydraulic pressure as a retarding force $(-\Delta P)$ and the osmotic pressure as the driving force $(-\Delta \pi)$. Of these membrane processes, only RO directly dewaters brines. FO and PAFO require a second process, most commonly a RO or thermal draw solute regeneration step, to produce a pure water permeate. Osmotically assisted reverse osmosis (OARO) (Figure 6.7), Hyrec proprietary, is a non-evaporative, membrane-based process that holds promise for high recovery, energy efficient desalination of high salinity brines [393].

High salinity feed is fed into an OARO module at a high hydraulic pressure. On the permeate-side of the module, the low-pressure sweep with a lower salinity flows counter-current to the feed. The high-pressure feed and low-pressure sweep establish a hydraulic pressure difference that is greater than the osmotic pressure difference across the membrane. The resulting water flux concentrates the feed and dilutes the sweep. The concentrated feed is the reject from the OARO process and may be crystallized or otherwise disposed of. If the diluted sweep concentration is relatively high and the target recovery is not achievable via RO, then the diluted sweep can be pressurized and fed into a second OARO module. Since the diluted sweep has a lower concentration than the original feed, an equivalent permeate volume can be realized with a lower sweep concentration. This second OARO module re-concentrates the diluted first sweep and dilutes the second sweep. The re-concentrated first sweep can be reused as the sweep inlet

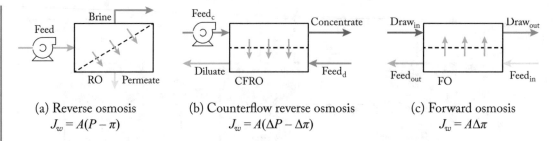

Figure 6.8: Diagrams of reverse osmosis, counterflow reverse osmosis, and forward osmosis stages with the equations governing water flux through their membranes. The color intensity of the blue arrows indicates the solute concentration of the stream, while the green arrows show the direction of permeate flow (reprinted with permission from [394]).

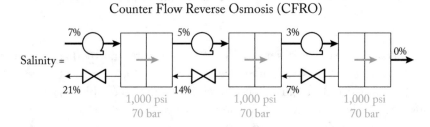

Figure 6.9: Apparatus of CFRO (reprinted with permission from [185]).

for the first OARO module and the diluted second sweep may require another OARO stage. The sweep concentrations successively decrease over a series of OARO stages until the sweep concentration is sufficiently low for RO. In the end, the OARO process will involve a feed inlet, a concentrated waste outlet, closed cycles of saline sweeps, and a product water outlet [393]. In April 2019, Saudi Arabia's Saline Water Conversion Corporation (SWCC) and Hyrec Technologies Ltd. have signed a Memorandum of Understanding to deploy Hyrec's Osmotically Assisted Reverse Osmosis (OARO) technology for Zero Liquid Discharge desalination in the Kingdom of Saudi Arabia.

In the Counter Flow RO (CFRO) process [185], Gradiant proprietary, dilute saline solution is introduced to the product side of the membrane to reduce the osmotic pressure barrier and thereby reduce the required feed pressure, which is essentially the same approach as the OARO, under a different name for marketing purposes. Brine is cascaded through multiple CFRO stages, concentrating it to saturation for disposal or crystallization, while producing a purified product water stream for beneficial reuse. Figure 6.8 compares single stages of RO, FO, and CFRO systems along with the equations that govern water flux through the membrane in each system.

Various configurations for CFRO have been discussed by Bouma et al. [394] (Figure 6.9). The split-feed configuration has the significant benefit of not having limitations on the maximum recovery ratio that are inherent to the configuration itself. Any limitations on recovery will come from other factors such as fouling and scaling, cost constraints, or the inability of materials to handle high salinity conditions. The additional degrees of freedom in the system allow for flexible operation at a wide variety of conditions. Split-feed CFRO shows the potential to operate very efficiently in a variety of regimes [394] (Figure 6.10).

Membranes constructed specifically for CFRO should seek to have high pure water permeability, a low structural parameter, and should tune salt permeability for maximum cost effectiveness. Purposefully developing "leaky" membranes may seem counterintuitive, but the reductions in ICP that result from salt passing through the membranes can significantly increase flux and allow for a system to operate with reduced membrane area, and thus reduced capital costs [394]. In an Indian textile plant, UF+NF+RO+SCE+CFRO was applied for wastewater treatment. An 82% recovery was expected, serving 127 m^3/d feed flow. In another case study, SCE+TMF+RO+CFRO was designed for 23 m^3/d feed, pursuing 95% recovery [185].

6.2.2   OSMOFLO-BRINE SQUEEZER

A dynamic membrane (DM), which is also called a secondary membrane, is formed on an underlying support material, e.g., a membrane, mesh, or a filter cloth, when the filtered solution contains suspended solid particles such as microbial cells and flocs. Organics and colloidal particles which normally result in fouling of the membrane will be entrapped in the biomass filtration layer, preventing fouling of the support material [395]. Formation of this cake layer, depicted in Figure 6.11, over the membrane surface can determine rejection properties of the system since the deposited layer will act as a "secondary" membrane prior the "real" membrane or support material [396].

Osmoflo Brine Squeezer (OBS) is able to operate at recoveries of up to 98% through innovative design, operation and membrane recovery [397]. Unlike other ROs, the Brine Squeezer operates at or above the scaling threshold of sparingly soluble salts, and concentrates feed water to a TDS of 100,000–120,000 mg/L. It recirculates concentrate in the high-pressure circuit to increase cross-flow and lower flux, and a thermally degradable coating, similar to the dynamic precoating, is applied to the membranes *in situ* to prevent irreversible fouling [398]. The membranes are fitted with spacers that are thicker than conventional RO systems to ensure removal of scaling compounds. The system is constantly monitored, and when scaling reaches a predetermined level, the unit is taken off-line, usually every one to four days, to undergo a high-temperature cleaning, before the membrane coating is reapplied. Brine Squeezers are designed in standard, 55 m^3/d modules, and systems are scalable to virtually any total production capacity, with a specific energy consumption that ranges from 6–8 kWh/m^3 (22.7–30.3/kgal), which is remarkably higher than the SEC of SWRO (10–15 kWh/kgal) [399]. Automation and instrumentation of OBS achieve operational autonomy, maximizes effectiveness of routine

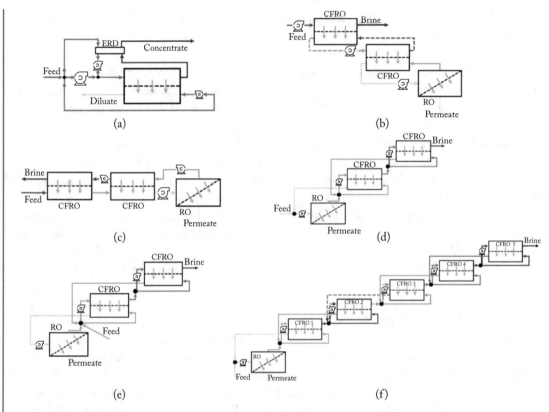

Figure 6.10: (a) Setup for a single split-feed CFRO stage, showing the high pressure and booster pumps to the left of the membrane, and the circulation pump to the right, as well as the energy recovery device (ERD), (b) simple consecutive loop configuration with two CFRO stages and an RO stage, (c) simple feed-through CFRO configuration with two CFRO stages and one RO stage, (d) a simple split-feed CFRO configuration with two CFRO stages and a single RO stage, (e) as a result of the additional degrees of freedom in a split-feed CFRO configuration, the feed stream may enter the system at some intermediate position in the system, as shown here, while still maintaining stable operation, and (f) 6-stage split-feed CFRO system. (Reprinted with permission from [394]).

cleaning functions, minimizes operator intervention and provides associated labor savings [397]. Table 6.1 shows the performance of an OBS system.

The OPEX of OBS varies from case to case. In the case of the CUB Yatala Brewery, the power consumption is 0.5 kW/m^3 and the OPEX is \$0.5/m^3. Maintenance CIP is 1–7 days, recovery CIP is 7–60 days. In the case of Leewood CSG produced water, the power consumption is 6–8 kW/m^3 and the OPEX is \$1.9/m^3. Weekly maintenance CIP is required. Recovery

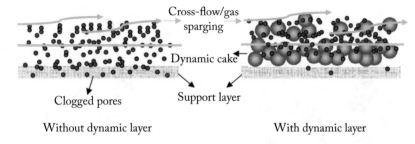

Figure 6.11: Demonstration of the dynamic cake layer (reprinted with permission from [395]).

Table 6.1: OBS performance [398]

Application	Problematic Constituents	Capacity (m³/day feed)	OBS Feed TDS (mg/L)	Improved Plant Recovery
Mine affected water	Heavy metals, transition metals, uranium, organics, bacteria	3,000	4,500–15,000	75 → 95%
Brewery wastewater	Hardness, silica, phosphate, organics, bacteria	1,200	4,000–7,000	70 → 85%
Coal seamgas produced water	Alkalinity, silica, barium, strontium, organics, bacteria	1,000	45,000–55,000	50 → >80%
Nickel refinery wastewater	Hardness, sulfate, nickel, cobalt	30 (pilot)	50,000–80,000	75% → >92%
Dairy whey minerals wastewater	Hardness, phosphate, organics	30 (pilot)	12,000–21,000	77% → >88%
Brewery–bore water	Hardness, silica, iron	6 (pilot)	18,000–21,000	70 → 92.5%

CIP is 3 months. In the case of the Ranger Uranium Mine, the power consumption is 2.5–3 kW/m³ and the OPEX is \$1.2/m³. Daily maintenance CIP is required. Recovery CIP is 6 to 8 weeks [398]. In contrast to the OPEX of typical BWRO even SWRO operation, the OPEX and maintenance of OBS are significantly higher. In return, the remarkable recovery and protection of the membranes still make it an attractive alternative.

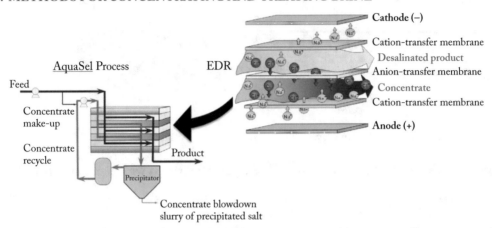

Figure 6.12: Diagram of the AquaSel process by GE. Brine is further concentrated using ED/EDR. The product is collected, while the ED concentrate is sent to a precipitator that removes calcium sulfate to prevent scaling. The concentrate is then recycled to the ED feed (reprinted with permission from [402]).

6.2.3   ELECTRODIALYSIS (ED) AND ELECTRODIALYSIS METATHESIS (EDM)

It has been discussed that ED can be applied to soften water. ED can also be applied to treat brines with high salinity. Wall et al. studied a pilot-scale EDR system, called AquaSel by GE, and its ability to concentrate brine from a brackish water RO facility [400]. An outline of the pilot system is shown in Figure 6.12. The brine was fed to the EDR system, and the concentrate from there was sent to a precipitator. The purpose of the precipitator was to collect and remove calcium sulfate from the concentrate to prevent it from scaling the ED membranes. The feed TDS concentration was around 6,500–7,500 mg/L. They were able to achieve 95% overall recovery of the brackish water, including the primary RO treatment (80% recovery of brine with EDR). A similar pilot study of EDR with a precipitator was also performed by Korngold et al. with brackish water RO brine. They were able to concentrate the 15,000 mg/L RO brine to 100,000 mg/L with an ED energy consumption of about 7.0–8.0 kWh/m^3 water produced [401]. For salinities up to 2,500 mg/L, the energy consumption typically ranges from 0.7–2.5 kWh/m^3. From 2,500–5,000 mg/L, the energy consumption for ED ranges from 2.6–5.5 kWh/m^3 [327]. As this trend suggests, electrical efficiency in ED decreases when concentration increases due to the lower yield of the electric fields [315].

The appeal of ED stems from its development at the industrial scale, its commercial availability, and that it can concentrate brine using only electrical energy [315]. This can be especially attractive to isolated areas that can use electrical energy sources like photovoltaic cells. ED is susceptible to scaling like other membrane processes are, but the extent of scaling in ED can be

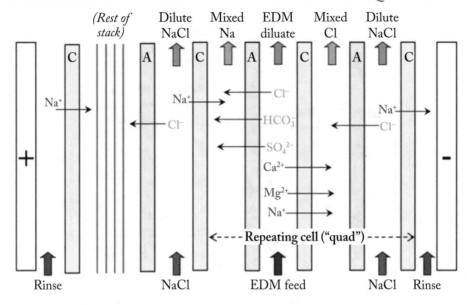

Figure 6.13: Diagram showing the repeating cell of an EDM stack and the movement of ions across anion-exchange and cation-exchange membranes. The resulting concentrate streams contain highly soluble salts (reprinted with permission from [403]).

reduced using periodic EDR operation. However, it is important to note that while ED excels at removing ions, it does not remove uncharged compounds from the feed waters [316]. As was mentioned earlier, ED also becomes less energy efficient at high salinities and cannot compete with other technologies like RO below a certain threshold [327]. At very high concentrations, ED becomes a competitive option because it is not limited by osmotic pressure like RO is.

Electrodialysis metathesis (EDM) is a new electrodialysis technology that can be used to further concentrate RO brine for subsequent salt recovery. As shown in Figure 6.13, EDM involves a series of ion-selective membranes between two electrodes. Each repeating unit consists of four membranes: one anion exchange, one cation exchange, one monovalent selective anion exchange, and one monovalent selective cation exchange membrane [404]. In each repeating unit there are also four solution compartments, two depleting and two concentrating [404]. While RO brine flows through one depleting solution compartment, a solution of NaCl flows through the other. They are arranged so that the RO brine feed flows between two concentrating compartments, and each concentrating compartment is between a NaCl compartment and an RO brine compartment [405].

Driven by an electrochemical potential difference between the anode and cathode, ions in the feed streams are transported across their respective membranes and switch ionic partners. As a result, two concentrated streams with highly soluble salts are produced: one containing sodium

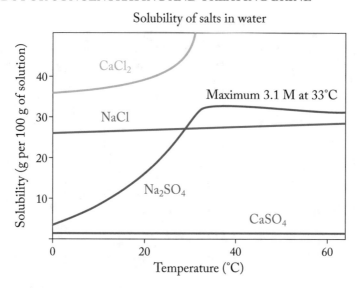

Figure 6.14: There is a significant difference in solubility between CaSO$_4$ and the salts that are present in the concentrate streams after EDM. This allows EDM to operate at high recovery without fear of membrane fouling (reprinted with permission from [406]).

with anions, and the other containing mostly chloride with cations. This process is advantageous because it reduces the precipitation of low-solubility salts such as CaSO$_4$ and CaCO$_3$, which cause scaling in RO and other membrane processes [404]. For this reason, EDM can achieve high water recoveries even when fed with RO brine. Figure 6.14 shows the large differences in solubility between CaSO$_4$ and the highly soluble salts present in the concentrate streams. After EDM, the two concentrate streams may be mixed together to produce CaSO$_4$ precipitates and a NaCl-rich supernatant, which can be reused as a NaCl solution feed for EDM [403].

Pilot and full-scale studies have been performed to research the technical and economic feasibility of EDM as an option for brine management, especially as an alternative to thermal brine concentrators for zero liquid discharge (ZLD). Bond et al. conducted a pilot study in which EDM further treated samples of concentrate from desalination plants [404]. They aimed for zero liquid discharge by treating the EDM concentrate streams with a crystallizer. The concentrate samples ranged from TDS 3,000 mg/L to 16,000 mg/L and were supersaturated with salts that would foul RO or ED membranes. Product water recoveries that ranged between 99.8 and 99.9% during the pilot tests were achieved. However, it was found that as the concentration of TDS of the EDM feed increased, EDM water recovery decreased and the energy required for EDM increased. Water recovery in EDM decreases because the amount of water in the EDM concentrate is proportional to the number of ions removed. Likewise, the energy required for EDM increases proportionally with the quantity of ions removed [404]. Note that the raw water

TDS refers to the TDS of water before it is fed to the initial RO desalination process. The typical energy required for a thermal brine concentrator is 20 kWh/m³ of RO concentrate treated and is mostly dependent on water volume rather than TDS concentration. It can be seen that the energy requirements for ZLD with EDM is below 20 kWh/m³ until about 15,000 mg/L raw water TDS. Based on their energy and cost analysis, the Bond et al. concluded that ZLD with EDM is very promising for brackish water sources containing up to 5,000 mg/L TDS, and not economically favorable for waters above 10,000 mg/L TDS [404].

Bond et al. recently conducted another full-scale EDM study in which they treated the brackish water RO concentrate from the Beverly Hills Water Treatment Facility [406]. The RO brine had an average 3,190 mg/L TDS. They were able to successfully treat the concentrate with 95% recovery and an energy consumption of about 0.6 kWh/m³. They also observed no increase in resistance due to membrane scaling in an EDM system over months of operation. Their research suggested that if EDM was used to treat all the plant's concentrate, the total plant recovery would increase from 74–95% [406].

According to these studies, EDM is a promising alternative for concentrate management and entertains the possibility of economical zero liquid discharge in the future. ZLD would first and foremost reduce disposal costs, especially for inland plants. EDM is a less costly and energy intensive method for ZLD compared to thermal desalination for moderately saline waters, which makes it appealing for inland areas that need to utilize non-traditional water sources. Because it separates RO brine into highly soluble salt solutions, EDM is less susceptible to fouling than other membrane processes. EDM also allows the recovery of valuable salts such as Na_2SO_4 and $CaSO_4$, which can offset some of the treatment costs [407]. However, while EDM technology presents many advantages over conventional methods, it still lacks the commercial support to develop the technology and bring it to market [406]. And, as is true with any water treatment technology, the performance and cost of EDM is dependent on the source water quality and treatment goals specific to that site.

6.2.4   MEMBRANE DISTILLATION (MD)

Membrane distillation is a thermally driven membrane separation process that can be used to treat saline waters [408] as shown in Figure 6.15. In MD, heated feed water is brought into contact with the surface of a hydrophobic, porous membrane [408]. On the other side of the membrane, cool feed water or coolant flows across a heat transfer surface to keep it at a low temperature. The cool and heated streams create a temperature difference across the membrane, and thus a vapor pressure gradient. As a result, water vapor from the feed permeates across the membrane toward the lower vapor pressure. The vapor then condenses on the heat exchange surface and is collected as product water. Under a threshold pressure, the hydrophobic membrane prevents both liquid water and salts from passing [408].

Depending on the method to collect the water vapor in the distillate side, MD can be classified into four basic configurations: direct contact membrane distillation (DCMD), air-

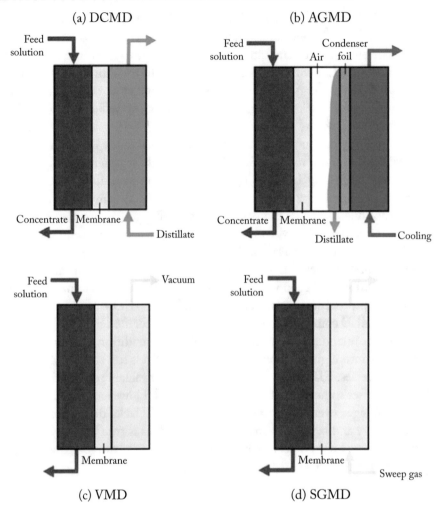

Figure 6.15: Schematic representation of the four basic MD configurations: (a) DCMD, (b) AGMD, (c) VMD, and (d) SGMD (reprinted with permission from [409]).

gap membrane distillation (AGMD), vacuum membrane distillation (VMD) and sweep gas membrane distillation (SGMD) [409]. In AGMD, the water vapor travels across an air gap before condensing on the cold heat transfer plate [315]. In VMD and vacuum-enhanced direct contact MD (VEDCMD), vacuum is applied on the permeate side of the membrane to further increase the vapor pressure gradient. In the direct contact versions of MD, the heated feed water is in contact with the active membrane surface, while a cold water stream is also in contact with the support side of the membrane [315].

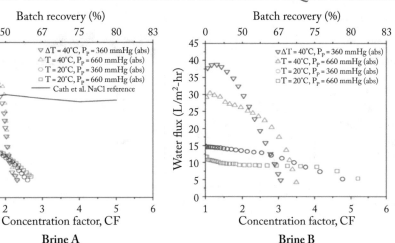

Figure 6.16: A comparison of MD runs with Brine A (left) and Brine B (right). The MD system is able to achieve higher recoveries with brine B despite its higher TDS concentration because it has a lower saturation level than brine A and contains residual anti-scalant (reprinted from [411]).

Martinetti et al. studied the performance of a bench scale VEDCMD system in treating brackish water reverse osmosis brines [410]. Brine A was collected from the RO concentrate from a groundwater desalting facility in Southern California and had an average TDS concentration of 7,500 mg/L. To generate brine B, brine A was chemically softened, separated from precipitated solids, treated with scale inhibitor, and finally further concentrated with RO. The average TDS concentration of brine B was 17,500 mg/L. The system was tested with two different temperature gradients (20 and 40°C) and two different vacuum pressures (360 and 660 mmHg). The permeate was at 20°C, while the feed was either heated to 40 or 60°C. A graph of the results of testing can be seen in Figure 6.16. Their results suggest that TDS concentrations within this range of feed concentration have little effect on initial water flux, as they are similar for both brines. This characteristic gives VEDCMD an advantage over pressure-driven membrane processes. It can also be seen that MD treatment of brine B resulted in much higher recoveries than for brine A. The researchers determined that this was due to the lower percent saturation values and the scale inhibitor present in brine B. Scale formation on the membrane surface was detrimental to water flux, as can be seen in the sudden drops in water flux in Figure 6.16, especially for the 40°C temperature difference runs. However, like FO, Martinetti et al. found that MD membrane cleaning allowed them to recover almost all of the initial water flux [410].

Lee et al. studied a bench scale DCMD system for brackish water RO brine treatment. In their study they compared the performance of MD, FO, and RO in treating the same feed

Feed Brine Composition

Water quality parameter	Concentration
Turbidity (NTU)	5.05
TOC (mg/L)	14.68
Hardness (ppm $CaCO_3$)	434.5
Ca (ppm $CaCO_3$)	311.2
Mg (ppm $CaCO_3$)	123.3
Na (mg/L)	141.5
Cl (mg/L)	305
SO_4 (mg/L)	360
K (mg/L)	36.5
SiO_2 (mg/L)	28.8

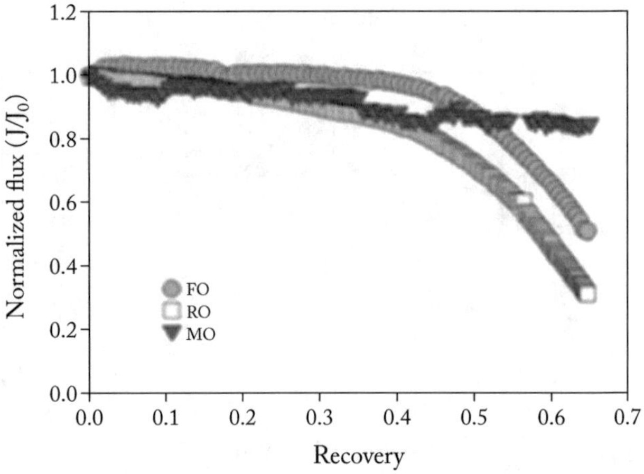

Figure 6.17: The composition of brine used as feed for the RO, FO, and MD systems (left). A graph comparing the change in relative flux against recovery for the three systems (right). According to Lee et al.'s study, MD seems to be less affected by membrane scaling, compared to FO and RO (reprinted with permission from [411]).

water [411]. The composition of the brine is shown in Figure 6.17 (left). The feed and permeate temperatures were 50 and 20°C, respectively. After the experimental runs, they found that while $CaCO_3$ scaling was present on all membranes (MD, FO, RO), the relative flux of MD was not affected as drastically as was seen for FO and RO. Both FO and RO relative fluxes begin to decline rapidly at around 50% recovery, while such a large drop was not seen even at 65% recovery with DCMD. This can be seen in the graph in Figure 6.17 (right). Lee et al. also tested the rejection of the RO, FO, and MD systems with NaCl solutions and found that all three rejected 99–100% of the salt [411].

Winter et al. studied PGMD with full-scale spiral wound modules. They investigated the effects of different feed flow rates and feed salinities on the specific energy requirements and distillate flow rates [412]. Synthetic salt water feed solutions were used at concentrations of 0, 13, 35, 50, 75, and 105 g/L, where the upper range of brackish water RO brine was around 75 g/L. For a feed solution with no salt (0 g/L), they found that as they increased the feed flow from 200–500 L/h, both permeate flow (10–25 L/h) and specific energy consumption, or SEC (130–207 kWh/m^3), increased with it. The SEC still increased despite the increase in permeate flow, because energy demand rose at a faster rate than the output flow did.

At constant temperatures (25°C and 80°C) and feed flow rates, they also found that the permeate production rate decreased with higher feed salinities. This change occurred because at a constant temperature, higher salt concentrations led to almost linear reductions in vapor pressure. While SEC increased with feed salinity under constant operating conditions, the re-

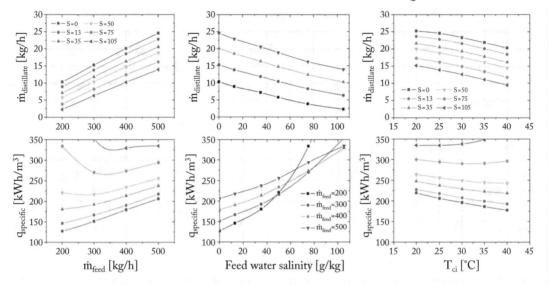

Figure 6.18: The effects of changing feed flow, feed salinity, and condenser temperature on SEC and permeate production in PGMD. Temperatures were held constant for the data shown in the left and middle columns at 25°C and 80°C. For the data in the right column, the feed flow was 500 kg/h and the evaporator temperature was 80°C (reprinted with permission from [412]).

lationship between SEC, salinity and feed flow rate was more complex and can be seen in Figure 6.18. At salinities under 50 g/L, lower feed flow rates were energetically advantageous, as was mentioned with fresh water feed. However, above 50 g/L, Winter et al. observed an energetic optimum that shifted to higher feed flow values with higher concentrations [410].

Finally, the researchers looked at the effects of changing the temperature differential across the membrane. For salinities below 75 g/L, SEC decreased with increasing temperature gradient. However, for salinities of about 75 g/L, greater temperature gradients become more energetically optimal. The SEC values that fit BWRO brine concentrations, 13–75 g/L, ranged from around 145–330 kWh/m³ and depended on the feed flow and the temperature differential [412].

MD can be operated at both high and low temperatures because the main driving force is vapor pressure differential across the membrane [413]. Therefore, the feed water does not need to be heated above the boiling point as long as a temperature differential is maintained. While energy requirements may exceed that of RO, they can be minimized when low-grade heat sources like waste heat from industrial processes are available to heat the feed water. MD processes are also able to produce very high quality distillate when the feed is composed of non-volatile compounds. However, in the presence of volatile organic compounds, there is a high likelihood of product water contamination [411]. The presence of organic compounds also increases the possibility of wetting the hydrophobic MD membrane, which would be detrimental to its performance. Conductive heat lost through the membrane is another drawback that must be addressed. The hydrophobic membranes used for MD, like PTFE, are typically more chemically inert than RO and NF membranes, allowing for more chemically aggressive cleaning. While the technology for MD is commercially available, research still needs to be done to improve the process.

6.2.5 DEWVAPORATION®

Dewvaporation is a thermal, humidification/dehumidification based process that enables additional water recovery from RO brine at temperatures below the boiling point of water (Figure 6.19) [408]. The tower consists of an evaporation side and a dew formation side separated by a heat transfer wall, usually made of thin plastic films [414]. On the evaporation side, saline feed water flows down through the top of the tower and wets the heat transfer (HT) wall. On the same side, a carrier gas, like air, flows from the bottom to the top of the tower. Heat is transferred through the HT wall to the evaporation side to heat the air and evaporate the feed water. As a result, hot saturated air leaves the top of the tower, while concentrated saline solution flows out the bottom [415]. Additional heat, typically waste heat, is added to the humid air before it is directed down the dew formation side of the tower. The heat is needed to establish a temperature difference between the two sides of the HT wall. On the slightly hotter dew formation side, the hot humid air begins to cool as water condenses on the HT wall. The resulting heat of condensation is transferred across the HT and acts as a heat supply to the evaporation side. Finally, product water condensate and the dehumidified air leave through the bottom of the tower.

Beckman et al. performed a pilot scale study of further wastewater RO brine treatment using dewvaporation in Phoenix, AZ [415]. The dewvaporation system was fed with 1,500 mg/L TDS RO concentrate and was able to concentrate the brine to 4,300 mg/L with a recovery of 78%. They found that for the runs with the highest thermal efficiencies, the heat required to produce 1,000 gallons of distillate was 764 kWh, or about 191 kWh/m^3 of distillate. If adequate waste heat is available, the energy requirements are reduced to the minimal electrical requirements of pumps and fans, about 0.5 kWh per 1,000 gallons of distillate, or 0.125 kWh/m^3.

In one run, the dewvaporation concentrate was recirculated to the feed, which eventually increased to 10,000 mg/L TDS. The system was still able to produce distillate with concentrations as low as 10 mg/L TDS and concentrate as high as 45,000 mg/L TDS. They concluded that increases in feed salinity did not sacrifice distillate quality, but decreased energy efficiency 30%. Altela, Inc. acquired the rights to the technology and has been conducting more pilot studies of dewvaporation for treating RO brine as well as produced water from oil and gas production [416].

There are several advantages to using dewvaporation as a method to recovery additional water from RO brine. Its main advantage over membrane processes is that mineral scaling poten-

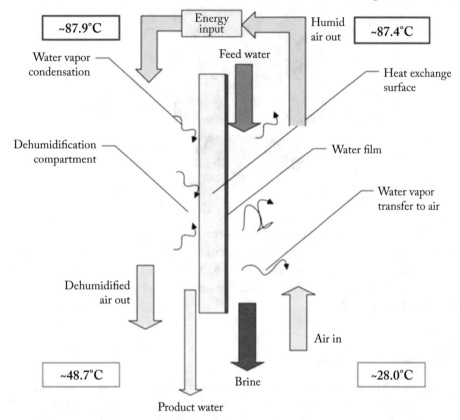

Figure 6.19: The humidification/dehumidification process used in dewvaporation. The evapo-ration side (right) and the dew formation side (left) are separated by a heat transfer wall. The temperatures in the diagram are typical temperatures presented by Beckman et al. (reprinted with permission from [415]).

tial is not a major problem, especially at high salinities [408]. This attribute makes it appealing for treating RO concentrates and produced water. The cost of equipment is minimal because dewvaporation operates at atmospheric pressure and typically utilizes inexpensive plastic sheets as the heat transfer wall [416]. It is an especially attractive technology when abundant waste heat is available, as it does not require water to be heated above its boiling point. However, there are still many drawbacks to dewvaporation. There is a high potential for product contamination when volatile organic compounds are present in the feed [414]. Research still needs to be done to improve the energy use efficiency of the system as well. In Beckman et al.'s study, they were only able to achieve about 1/13 of the theoretical thermal efficiency. Most notably, the energy and heat demand is very high, making it unfeasible when waste heat is not available.

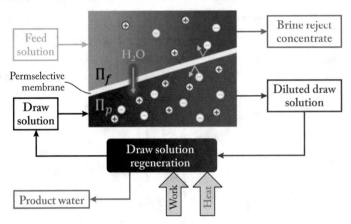

Figure 6.20: Diagram of the forward osmosis process. Driven by an osmotic pressure gradient, water flows from a lower salinity feed to the higher salinity draw solution. The feed side becomes more concentrated, while the draw solution is diluted. The dilute draw solution is then treated to both recover product water and to regenerate the high concentration draw solution (reprinted with permission from [417]).

6.2.6   FORWARD OSMOSIS (FO) MBC

Forward osmosis is a membrane process in which water flows from a relatively low concentration feed solution to a more concentrated draw solution [417]. The general process of forward osmosis can be seen in Figure 6.20. Rather than hydraulic pressure, the driving force for the flow of water across the membrane is an osmotic pressure gradient. Similar to RO membranes, the semi-permeable FO membrane allows water to flow through, but acts as a barrier to particulate matter and most dissolved constituents [414]. As a result, the process concentrates the feed while diluting the draw solution, which is usually composed of NaCl or other salts. Some combination of heat and work is then required to separate the product water from the reusable high concentration draw solution [417].

In many cases, another RO system is used with FO to re-concentrate and reuse the draw solution. Martinetti et al. conducted a bench scale study in which a FO-RO system was used to recover additional water from a brackish water RO concentrate [410]. The brine solutions fed to the system were 7,500 mg/L and 17,500 mg/L TDS, and the draw solution was a 50 g/L NaCl solution. During the FO, they achieved recoveries of 86 and 60% for the two feeds, respectively. When considering the 70% recovery of the initial RO process, the total recovery after FO was around 95% for both feeds. Even though high recoveries can be achieved with an FO-RO system, the nature of the process makes it inherently more energy consumptive than stand alone RO [418]. Research about novel draw solutes and regeneration technologies has been carried out to try to solve this issue, but energy savings are limited by the minimum energy of separation. Moreover, the energy requirement of state-of-the-art RO processes already approaches this limit [418]. It is not likely that any FO technology will outperform RO in this area, so when it can be used, RO will remain the more energetically favorable option [418].

One reason FO is still an attractive technology is that membrane fouling is not intensified by hydraulic pressure in FO, as it is in RO. While the cake layer formed on the membrane in RO is densely compacted, the cake layer formed in FO is thicker but much more loosely packed [418]. Therefore, irreversible flux is minimal and foulants can be removed easily with backwashing, chemical cleaning, or increasing cross-flow velocity [408]. Typically, as much as 80–100% of the initial FO water flux may be recovered through periodic rinses of the membrane surface [418]. FO can therefore be used when feed salinities have high fouling potential and high osmotic pressures that exceed the operating pressure of RO [418]. In these cases, when RO cannot be used to regenerate the draw solution, vaporization of thermolytic draw solutions may be an alternative.

McGinnis et al. performed a pilot study of an FO membrane brine concentrator (MBC) for desalination of high salinity brines using a NH_3/CO_2 draw solution [419]. This draw solution contained a mixture of thermolytic ionic solutes that vaporize when heated. The feed consisted of produced water from natural gas production in the Marcellus shale and had an average concentration of 73,000 mg/L TDS. A diagram of the FO desalination process is shown in Figure 6.21. As can be seen, after water extraction from the draw solution, the product water is sent to RO to meet quality standards. The pilot study resulted in a recovery of about 64% recovery, brine concentration of 180,000 mg/L TDS, and product water concentration of 300 mg/L TDS. The thermal energy required for the FO MBC system was about 275 kWh/m^3 of product water. As a comparison, an evaporative brine concentrator evaporating 73,000 mg/L NaCl water would have had a thermal energy requirement of about 633 kWh/m^3.

Using FO for additional water recovery from RO brine is most feasible when the brine concentrations are above the threshold for RO use. FO's utility lies in its ability to treat very high salinity waters with high fouling potential. In these cases, thermolytic draw solutions can be used to reduce the thermal energy requirements compared to conventional thermal brine evaporators, although the energy requirements still remain high due to the high salinity of the feed. The use of FO and thermolytic draw solutions can be especially economical when low-cost or waste heat is available for the thermolytic draw solution regeneration process [418]. However, as mentioned earlier, when RO can be used, it is unlikely that FO will be more energetically favorable.

6.3   BRINE POST-TREATMENTS FOR ZERO LIQUID DISCHARGE

Zero liquid discharge (ZLD), an ambitious desalination management strategy that eliminates liquid waste and maximizes water recovery, has attracted renewed interest worldwide in recent years. Normally, ZLD can be achieved by placing crystallizers or evaporation ponds at the end of the brine disposal process, since the residual water will be "squeezed" out from the brine with

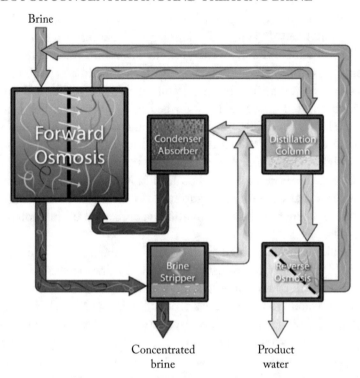

Figure 6.21: A diagram of the pilot FO membrane brine concentrator system used by McGinnis et al. to treat very high salinity produced water. The draw solution consists of thermolytic solutes that can be separated from water using low-grade heat. The water is treated by RO last to meet quality requirements (reprinted with permission from [419]).

very high TDS via such processes, where a crystallizer is more popular. Evaporation ponds can be used as competitive alternatives to brine crystallizers. Nevertheless, they are only suitable when treating small volumes at locations with a high evaporation rate and inexpensive land. Their high capital cost and environmental concerns for potential leakage of hazardous waste further hinder widespread application. Although implementation of ZLD reduces water pollution and augments water supply, the technology is constrained by high cost and intensive energy consumption [420]. A typical brine concentrator and crystallizer are shown in Figures 6.22–6.24, they serve as potent technologies to squeeze the water out of the brine, at the cost of significant energy consumption, and which therefore are frequently designed and applied to ZLD/MLD operations.

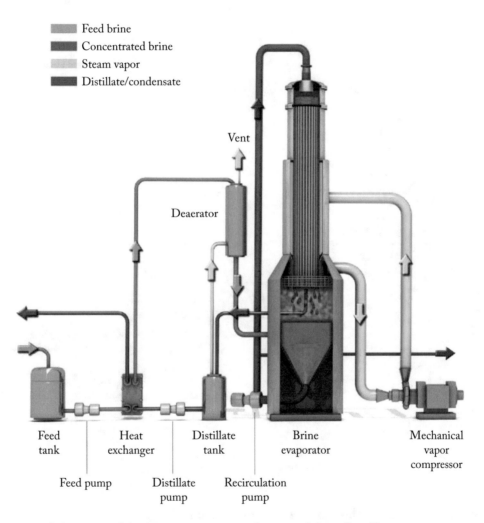

Figure 6.22: Schematic of the brine concentrator (reprinted from [183]).

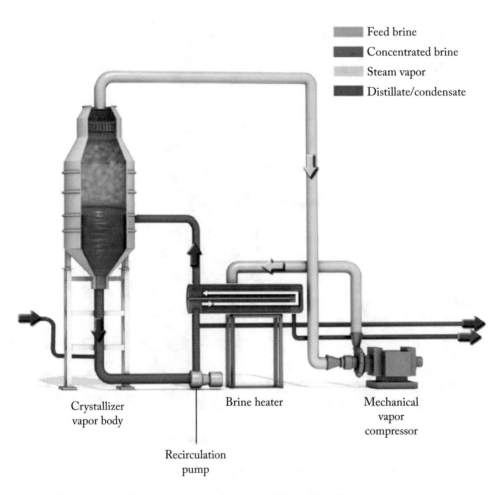

Figure 6.23: Schematic of the crystallizer (reprinted from [183]).

Figure 6.24: Brine concentrators and crystallizer in an industrial plant (reprinted from [183]).

6.3.1 EVAPORATIVE BRINE CONCENTRATION (EBC) METHODS + CRYSTALLIZER

Evaporative Brine Concentration (EBC) methods can be coupled with a crystallizer to achieve ZLD. The concentrated brines produced by brine concentrators, normally MVC, are usually sent to a brine crystallizer where the remaining water is further recovered. Similar to brine concentrators, vapor compressors are employed in crystallizers to supply the heat needed for water evaporation. However, for small systems (less than 23 L/min), steam-driven crystallizers are economically favorable. Vapor compressor crystallizers are commonly operated in a forced-circulation mode. The viscous brine is pumped through submerged heat exchanger tubes under pressure, thereby preventing boiling and subsequent scaling inside the tubes. The energy consumption of crystallizers is as high as 62–66 kWh/m^3 of treated feedwater, which is nearly three times that consumed by MVC brine concentrators, shown in Figure 6.25. This dramatic energy increase is inevitable as crystallizers are treating feed brines with much higher salinity and viscosity [420].

A typical configuration for achieving ZLD using an MVC system is provided in Figure 6.26. The specific consumption of these facilities is lower than for other distillation processes, with the equivalent electrical consumption normally about 10 kWh/m^3 [421]. The limitation of this type of technology is the maximum size of the volumetric compressors used. Its maximum capacity does not allow for high production of desalinated water.

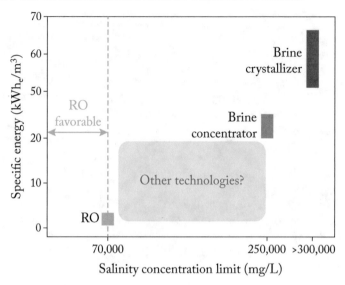

Figure 6.25: Specific energy consumption by an RO brine concentrator, and brine crystallizer. Although RO is energy efficient, its limited salinity range (typically with an upper concentration of ~70,000 mg/L) provides opportunities for other technologies to be applied in ZLD systems. The specific energies shown in the figure are in kWh per cubic meter of feedwater (reprinted with permission from [420]).

6.3.2   MEMBRANE-BRINE CONCENTRATION (MBC) TECHNOLOGIES + CRYSTALLIZER

Similarly, Membrane-Brine Concentration (MBC) technologies, OARO, FO and ED for instance, even though effective at concentrating brine, are still not capable of achieving ZLD without combining with crystallizers. The world's first FO-based ZLD system was constructed at the Changxing power plant in Zhejiang Province, China. The system treats a mixture of flue gas desulfurization wastewater and cooling tower blowdown at 650 m³/day. The feedwater is first concentrated by RO to a concentration of ~60,000 mg/L. The NH_3/CO_2 FO process is then used as a brine concentrator to further concentrate the RO brine to above 220,000 mg/L TDS. As a final step, the FO brine is fed to a crystallizer for further concentration, while a high-quality product water (TDS < 100 mg/L after polishing by a secondary RO) is produced for reuse as boiler makeup water [420].

Oren et al. [422] demonstrated a pilot RO-EDR system for brackish water desalination with a water recovery of 97–98%. In the system, an EDR concentrated the RO brine to a salinity of 100,000–200,000 mg/L prior to a side-loop crystallizer and wind-aided intensified evaporation. Loganathan et al. [423] studied another EDR-ZLD system. EDR effectively removed hardness to reduce the scaling potential of saline basal aquifer water, thereby improving the

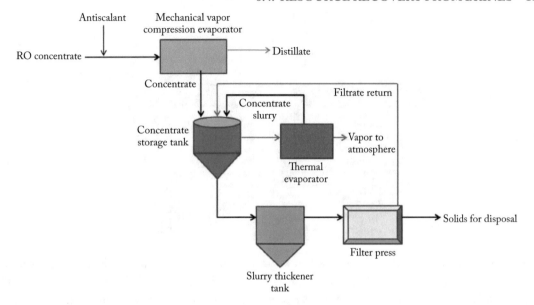

Figure 6.26: Process schematic of ZLD process using mechanical vapor compression and thermal evaporator (reprinted with permission from [384]).

subsequent RO recovery without chemical addition. The EDR brine could reach a salinity of 125,000 mg/L and was further concentrated by a brine crystallizer to approach ZLD.

A conceptual near-ZLD system incorporating MD with reverse electrodialysis (RED) was shown to achieve both water and energy production in seawater desalination. In that system, MD reduced the volume of simulated SWRO brine (1 M NaCl) by more than 80%. The produced MD brine was then mixed with seawater in a RED stack to generate electrochemical energy [424]. However, large-scale applications of MD are still hindered by its technical immaturity and low single-pass, single-module water recovery [420].

6.4   RESOURCE RECOVERY FROM BRINES

6.4.1   SALINITY GRADIENT SOLAR PONDS

Salinity gradient solar ponds (SGSP) use salinity gradients to trap and store solar energy in the deepest region of the pond [414]. As can be seen in Figure 6.27, these ponds consist of three zones—each more concentrated and dense than the one above it. The uppermost region, called the surface zone or the upper convective zone (UCZ), is a homogeneous layer of low salinity water, typically low-salinity brine or fresh water. Below the UCZ, the main gradient zone, or non-convective zone (NCZ), has increasing salinity and density with depth. As its name suggests, this zone does not experience convection and serves as a transparent insulating layer.

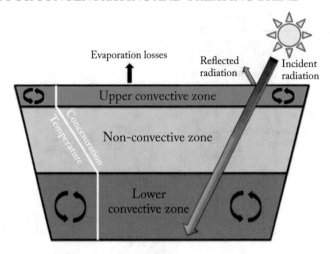

Figure 6.27: Diagram of a salinity gradient solar pond and its three layers. Solar energy is able to penetrate through the UCZ and the NCZ to be stored in the LCZ. Unlike the homogeneous UCZ and LCZ layers, the NCZ contains a salinity gradient and a similar temperature gradient (reprinted with permission from [425]).

Finally, the bottom zone, known as the lower convective zone (LCZ) is a homogeneous region of highly concentrated salt solution. Solar energy is collected and stored in this region, resulting in a temperature increase [414]. Studies have shown that the LCZ can reach temperatures up to 110°C, about 70°C higher than the UCZ temperature [425]. The high heat capacity and large volume of water along with the insulating NCZ make salinity gradient solar ponds both solar thermal collectors and long-term thermal storage devices [426].

In some areas, especially in hot and arid regions, salinity gradient solar ponds are appealing options for brine management and disposal. In this method, brine can be used to fill the high salinity LCZ in solar ponds. Through thermal energy capture, the brine may then be converted to clean energy to power processes such as desalination [426]. However, similar to the other energy recovery methods, salinity gradient solar ponds may not be a feasible brine management method for inland areas with very limited access to fresh water. This is because evaporation occurs at the UCZ surface and must be frequently replenished with fresh water or low salinity water. When freshwater production is the main objective, it is not advantageous to have to use the product water in the solar pond ([182], Ruskowitz, 2013). In addition, the ponds require large areas of land, and heat recovery from low salinity brackish water brine is not feasible in solar ponds.

6.4.2   SALINITY GRADIENT POWER – PRO, RED, MEB

There is a large chemical energy gradient between high salinity solutions (seawater, brackish water, RO concentrate) and low salinity solutions [414]. This chemical energy is released when the solutions are mixed together. Salinity gradient power recovery methods, such as reverse electrodialysis (RED), pressure-retarded osmosis (PRO) and mixed entropy batteries (MEB), capture this energy and convert it to useful electricity.

RED and PRO both utilize semi-permeable membranes to generate energy. RED uses ion-selective membranes to allow ion, but not water, transport through the membrane [427]. RED consists of a stack of alternating cation-exchange membranes (CEM) and anion-exchange membranes (AEM) between an anode and a cathode. Cations pass through the CEMs and anodes pass through the AEMs. The channels contained within each membrane pair are fed with an alternating pattern of high and low salinity solutions, creating salinity gradients across each membrane. While ions flow from high salinity to low salinity, positive and negative ions flow in opposite directions due to the ion-selective membranes. The resulting difference in electrochemical potential creates an electrical current at the electrodes [427].

PRO uses water-permeable membranes to produce pressurized water, which in turn helps to generate electricity [427]. In PRO, two chambers are separated by a semipermeable membrane. A low concentration feed flows through one chamber, while a high concentration draw solution flows through the adjacent chamber. Driven by the salinity gradient, water from the low concentration side flows across the membrane to dilute the high concentration solution, increasing the volumetric flow rate of that side. The resulting high-pressure solution can then drive a hydro-turbine to generate energy. A portion of the high-pressure solution is also sent back to a pressure exchanger that uses its pressure energy to pressurize the low-pressure high salinity feed [427].

Rather than membranes, MEB uses electrodes to store and collect chemical energy. The device consists of an anionic electrode, which selectively interacts with Cl^- ions, and a cationic electrode, which selectively interacts with Na^+ ions. The MEB process is outlined in Figure 6.28. First the electrodes are submerged in a low salinity solution in their discharged states, when the appropriate ions are incorporated into the electrodes' structures. In this step, the dilute solution removes ions from the electrodes, thus charging the battery [428]. Next, the dilute solution is replaced by a concentrated solution to increase the potential difference between the electrodes [428]. The high potential difference reincorporates anions and cations into their respective electrodes, discharging the battery [428]. Finally, the concentrated solution is exchanged back with a dilute solution that decreases the potential difference between the electrodes, and the cycle begins again.

In all three of the mentioned salinity gradient power generation methods, seawater (or brine) and fresh river water are commonly used as the high salinity and low salinity solutions, respectively. While brackish water RO brine can be an appropriate substitute for the high salinity solution, many inland areas do not have access to the low salinity water necessary to use these

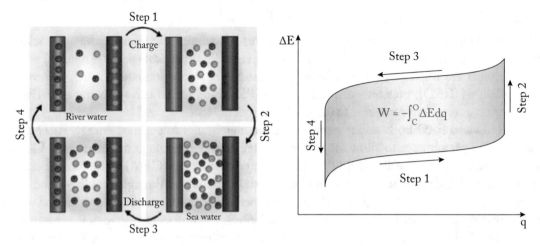

Figure 6.28: A schematic outlining the steps of MEB use (left). The cycle of battery cell voltage (ΔE) vs. charge (q) (right) shows that no energy is consumed or produced in Steps 2 and 4. Step 1 requires energy to charge the battery by removing ions from the electrodes and Step 3 produces energy to do the opposite (reprinted from with permission [428]).

methods. Furthermore, after energy recovery, the saline solutions will still need to be disposed of in some way. Even in areas that have the resources to operate PRO, RED, or MEB, many argue that substantial research is still needed before it can become economically feasible. For example, research must be done to improve energy conversion efficiencies, to minimize fouling and scaling on membranes and electrodes, and to evaluate the cost effectiveness [414]. While energy recovery appears to be an attractive beneficial use of brine, it is not yet feasible for the inland brine management.

6.4.3   SELECTIVE SALT CRYSTALLIZERS

The chemical components in the concentrate can be solidified and recovered for additional applications. It should be noted that the goal of these applications is to extract salts; water may not be recovered through the physical and chemical processes [410]. Converting concentrate from a waste to a resource through treatment and beneficial use may minimize both costs and environmental impacts. Mekorot Water Company owns and operates a dual purpose SWRO plant in Eilat that produces desalinated water and high-quality table salt [429]. The feed to the desalination plant is a blend of 80% seawater and 20% BWRO concentrate from adjacent BWRO plants. The concentrate from the SWRO plant is blended with seawater, and this stream is fed to a series of evaporation ponds, and thereafter to the salt processing factory of the salt company [410].

Saltworks Technologies has also partnered with SPX Cooling Technologies to develop an air-cooled tower that is able to achieve ZLD and harvest the precipitated salts [410]. Since the solubility of target products, i.e., salts such as $NaCl$, Na_2CO_3, and $NaSO_4$, is highly dependent on the temperature of the solution or brine, the selective crystallizers are able to recover and produce various salts from the brines via multiple temperature-control operations. The diversity and purity of the recovered salts are related to the brine source.

EWM is finalizing commissioning on its first commercial plant in El Paso, Texas, using Full Recovery Desalination^{SM}, which extracts the dissolved solids contained in a variety of water sources for beneficial use, leaving potable water for municipal and other uses [430]. The plant is located adjacent to El Paso Water Utilities' Kay Baily Hutchison (KBH) Water Treatment Plant. It is designed to process waste brine concentrate currently being disposed from the KBH plant through deep well injection approximately 20 miles from the site. In addition, the plant will process raw brackish well water. It's expected to produce ~2.4 million gallons per day (mgd) potable-quality water from 1.3 mgd of raw brackish water and 1.3 mgd concentrate, with influent raw brackish water TDS: 2,500 mg/L and influent concentrate TDS: 13,000 mg/L. The products will be (1) potable-quality water (tds < 800 mg/l), (2) caustic soda (50% concentration), (3) hydrochloric acid (35% concentration), (4) gypsum (high purity, 100% soluble), and (5) magnesium hydroxide (98% purity, 56% solid).

Veolia Water Solutions & Technologies [431] patented a process that provides for recovering sodium chloride crystals and sodium carbonate decahydrate crystals from a concentrated brine that results from a gas mining operation where gas and produced water is recovered and the produced water constitutes the brine. An initial pre-concentration process is carried out where the brine is concentrated and in the process carbon dioxide is removed from the brine and at least some sodium bicarbonate is converted to sodium carbonate. In one process, the concentrated brine is directed to a sodium chloride crystallizer where the brine is heated and further concentrated to form sodium chloride crystals that are separated from the brine to yield a product and wherein the resulting brine is termed a first mother liquor. The first mother liquor is then directed to a sodium carbonate decahydrate crystallizer where the first mother liquor is cooled and concentrated resulting in the formation of sodium carbonate decahydrate crystals and a second mother liquor. The second mother liquor is split into two streams where one stream is directed back to the sodium chloride crystallizer, while the other stream is wasted or further treated (Figure 6.29).

6.4.4   BIPOLAR MEMBRANE ELECTRODIALYSIS (BMED) FOR HCL AND NAOH PRODUCTION

BMED is a version of electrodialysis that can be used to produce acids and bases from RO brine [432]. Figure 6.30 shows a schematic of the process. BMED involves bipolar membranes (BPM) as well as cation and anion exchange membranes between two electrodes. A BPM typically consists of three layers: a strong acid CEM, a weak base layer, and a strong base AEM.

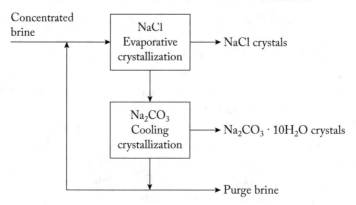

Figure 6.29: Selective salt recovery from a mixed salt brine.

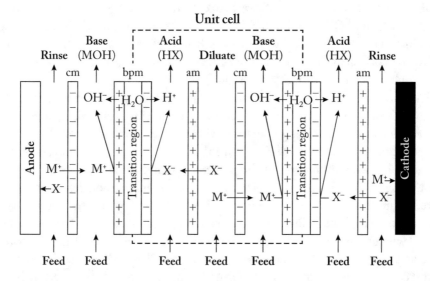

Figure 6.30: Schematic of bipolar membrane electrodialysis. Bipolar membranes allow the splitting of water and separation of the resulting H^+ and OH^- ions. The ion exchange membranes are arranged so that H^+ ions and anions from the feed mix to form acids and OH^- ions mix with cations to form bases (reprinted with permission from [433]).

The weak base layer in the middle promotes water splitting into H^+ and OH^- [433]. When an electric field is applied, anions flow through the AEMs toward the anode and are stopped by CEMs, and the opposite occurs for cation movement. As a result, hydrogen ions and anions from the saline brine are trapped together to produce acids, and hydroxide ions and cations from the feed are combined to form bases [407, 433].

Table 6.2: NaCl concentrations in the feed solution (C) and maximum acid and base concentrations achieved with fraction of their maximum possible concentration [433]

C (mM)	Acid (mM)	Base (mM)
48	39/81%	37/77%
95	73/77%	69/73%
192	148/77%	144/75%
390	303/78%	299/77%

Table 6.3: Cost comparison for purchasing the chemicals in bulk compared with the energy cost for onsite production, assuming electrical energy is priced at $0.07/kWh [433]

Chemical	Purchase cost ($ ton-1)	Purchase cost ($ kmol-1)	BMED cost ($ kmol-1)
35% HCl	320	37	3.5–12.6
99% NaOH	525	21	3.5–12.6

Note: A freight cost of $100/ton was added to the purchased reagents. The BMED energy cost includes the electrical energy to run the BMED cell and electrical pumping cost, which is assumed equivalent to the cost to operate the cell.

Davis et al. performed a lab scale study of BMED in which they fed the system with salt solutions between 2.8 g/L and 22.8 g/L NaCl [433]. The system consisted of 8 repeating units of a CEM, BPM, and AEM. After a single pass, the BMED system was able to produce acids and bases with molar concentrations of about 75% of the feed salt concentration. Current utilizations in these experiments were greater than 75%, increasing with higher feed salinities. At the same time, the energy required per unit of acid or base produced increased with increasing current density. They found that the energies required for acid production with their feed concentrations were 0.55–2.47 kWh/kg of acid produced. From their economic analysis, Davis et al. found that the total cost of acid and base production (including energy costs, equipment costs, cost of capital, and cost of membrane replacements) remained cost competitive with buying acids and bases [433] (Tables 6.2 and 6.3).

Another lab scale study of acid and base production with BMED was performed by Yang et al. [434]. Their feed consisted of 40 g/L of NaCl, which is on the higher end of brackish water brine concentrations. The water was softened prior to BMED to prevent scaling on the membranes, so membrane scaling was minimal even during long term testing. Their system was able to produce a maximum acid concentration of 1.2 M in batch mode, and an acid concentration of 1 M when operating in continuous mode. They also confirmed that the product water, or diluate, contained only low levels of major ions. The current utilization in their system was observed

Table 6.4: The compositions of the produced acid, base, and product water [434]

Composition (mg/L)	Acid	Base	Product
Sodium	106	9400	196
Potassium	5.3	400	10.2
Calicum	0	1	0
Magnesium	0	0	0
Chloride	23000	123	221
Sulfate	3700	18	21
Boron	0.8	0.2	036
Silicon	2.1	2.1	0.6
Bromide	103	5	0.7
DOC	0.1	0.7	0.1

between 52–74%, and the specific energy requirements were between 7.5–8.2 kWh/kg of acid produced. The energy consumption of BMED for producing acids and bases from desalination and industrial brines is typically between 2.3 and 18 kWh/kg of product [435] (Table 6.4).

The major advantage of using BMED to treat RO brine is that it converts the high salinity brine into high-demand chemicals that are easy to sell, thus minimizing or eliminating the need for brine disposal [407]. It is especially advantageous when the acid or base products can be reused in RO feed or RO brine treatment onsite, instead of being purchased from a location further away. However, the technology still has several limitations that must be overcome.

The technical barriers are related to imperfect permselectivity and the effect of electro-osmosis [435]. The flaws in the permselectivity of ion exchange membranes cause the migration of co-ions and thus impair the purity of the products. The selectivity of both monopolar and bipolar membranes also decrease with increasing salinity and acid/bath concentrations, respectively, leading to contamination of the products. Electro-osmosis causes water leaks through the membranes and so limits the production of high concentration acids and bases using BMED [435]. Currently, it is difficult to maintain high-energy efficiencies in BMED systems that involve high concentrations and high current densities due to ion leakage between membranes. Another factor limiting the large-scale use of BMED is the high cost of electrodes and membranes, especially bipolar membranes. Bipolar membranes can be 3–10 times more expensive than monopolar membranes because of the complicated manufacturing process and the very limited choice of manufacturers.

The presence of impurities in the feed affects the purity of the resulting products unless additional treatment steps are added. As mentioned earlier, if acids and bases are needed in the desalination process, the products of BMED can be used onsite. In this case, impurities in the

Table 6.5: Brine treatment methods and their advantages and disadvantages

Treatment Method	Level of Testing	Tested Feed TDS (mg/L)	SEC of Given Feed TDS (kWh/m³ of water treated)	Benefits	Limitations
WAIV	Pilot	12,000–150,000	N/A	Much higher evaporation rates than evaporation ponds; possible salt recovery	Land intensive; possible contamination of groundwater
Phyto-desalination	Pilot	2,600–3,200	N/A	Beneficial/green reuse of brine; halophytes can be grown and used for grazing, landscaping, habitat development, etc.	Land intensive; possible contamination of groundwater; buildup of salt in soil
Bipolar Membrane Electrodialysis	Lab	40,000	7.5–8.2 kWh/kg of acid produced	Produces acids and bases that can be sold	Membranes are expensive and complicated to make; very few manufacturers of BPMs;
Electrodialysis Metathesis	Pilot/full	3,190	0.6 (electrical)	Allows recovery of valuable salts that can be sold; high resistance to scaling	Lacks commercial support to develop the technology
Dewvaporation	Pilot	1,500	191 (0.125 kWh/m³ for electrical requirements)	Resistance to scaling; low cost of equipment; can utilize low-grade heat	High energy demand; considerable footprint; high potential for contamination when VOCs are present
Forward Osmosis	Pilot	180,000	275	Irreversible fouling is minimal; able to treat high salinity waters that exceed working pressure of RO	When RO can be applied, it is unlikely that FO will be more energetically favorable
Membrane Distillation	Pilot	13,000 (with feed flow 200 kg/h)	150	Can utilize low-grade heat; less affected by scaling than FO or RO	High energy demand; high potential for contamination when VOCs are present
Electrodialysis w/ Reversal	Pilot	15,000	7.0–8.0 (electrical)	Technology is developed at the industrial scale; reversal operation minimizes scaling; low energy requirements	Only removes ions while uncharged species remain; becomes less energy efficient at high salinities

acids and bases are acceptable because the feed already contains them [435]. Before BMED can be used for wide scale applications, research must be done to address the previously mentioned limitations.

6.4.5   BMED FOR CO_2 MINERALIZATION WITH CALCIUM

Carbon dioxide mineralization with calcium has been evaluated on the lab scale. Zhao et al. [436] designed and evaluated the feasibility of using BMED coupled with a crystallizer for CO_2 mineralization. Normally, a mineral carbonation process should proceed under a reactively high solution pH. In this system, a four-chamber bipolar membrane electrodialysis is introduced, in which CO_2 is adsorbed and ionized in the alkaline chamber. Then carbonate ions migrate to the salt chamber, where they combine with seawater, but are controlled in the ionic state to prevent membrane contamination. Seawater or brine is directed into a crystallizer where calcium carbonate nucleates and precipites.

CHAPTER 7

Renewable-Powered Desalination

7.1 TYPES OF RENEWABLE ENERGY FOR DESALINATION

Growing requirements of freshwater and unsustainable nature of fossil fuels are driving the interest in using renewable energy for desalination applications [15, 437–439]. Renewable energy is derived from natural processes that are replenished constantly (electricity and heat generated from solar, wind, ocean, hydropower, biomass, geothermal resources, biofuels and hydrogen derived from renewable resources).

Various sources of renewable energy, shown in Figure 7.1, are available in different regions across the globe. Main conventional renewable energy sources of interest for desalination include solar, geothermal, wind and tidal wave. Besides the conventional renewable energy sources, salinity gradient power (SGP), also termed as Blue energy, represents a green and nearly unexplored source of sustainable energy. Blue energy can be harnessed by controlled mixing of two liquid streams with different salinity levels and has great potential in alleviating the energy demand and carbon footprint of desalination plants.

7.1.1 SOLAR ENERGY

Solar energy can be harnessed directly as electricity, or as solar thermal energy, which is either used in heating or cooling systems, or drives turbines to generate electricity. Technologies for solar energy therefore falls under two broad categories: PV and solar thermal (Figure 7.2). Solar thermal technologies are further divided into concentrated solar power (CSP), for electricity generation, or direct use in low-temperature heating applications [441].

Solar energy is the most abundant form of renewable energy across the globe. It has been reported that many regions in Middle East and North Africa receive 5–7 kWh of solar insolation each solar day. Most of these regions are rich in brackish or seawater but suffer from lack of sufficient freshwater, making them ideal for solar energy drive desalination. Modern solar desalination plants generate steam by using solar energy that runs desalination unit mechanically. Due to its abundant availability and flexibility of converting into electric or thermal response, solar energy has been the main focus for renewable desalination [440].

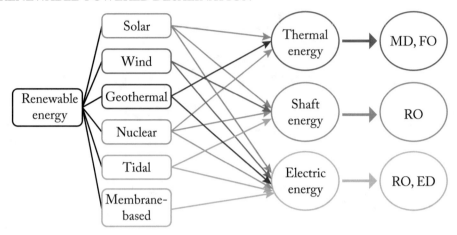

Figure 7.1: An overview of possible renewable energy technologies for various membrane-based desalination techniques (reprinted from [440] with permission).

7.1.2   GEOTHERMAL ENERGY

Geothermal energy is the form of renewable energy stored in the earth that can be pumped in form of steam and hot water which can further be used to generate electricity. There are different geothermal energy sources. They may be classified in terms of the measured temperature as low (below 100°C), medium (100–150°C) and high temperature (above 150°C). The thermal gradient in the Earth varies between 15°C and 75°C per kilometer depth nevertheless, the heat flux is anomalous in different continental area [442]. Depending upon the quality of geothermal energy, it can be used either directly in the form of steam to run a turbine or to vaporize a low boiling point fluid that can be further used to run the turbine.

Geothermal energy is a proven technology for electricity production, although not spread out commercially. By 2010, more than 10,000 megawatts energy was supplied by geothermal energy in 24 different countries across the globe, serving more than 60 million people [443]. Many regions with good resources of geothermal energy face the water shortage problem. A high-pressure geothermal source allows the direct use of shaft power on mechanically driven desalination, whereas high-temperature geothermal fluids can be used to generate electricity to drive RO or ED plants. In case of low temperature geothermal energy source, the heat can be directly used for thermal desalination. As compared to the other renewable energy resources, geothermal resource offers the uninterrupted thermal energy [440].

7.1.3   WIND ENERGY

Wind is generated by atmospheric pressure differences, driven by solar power [442]. Wind energy is an abundantly available source of renewable energy around the world. Coastal areas,

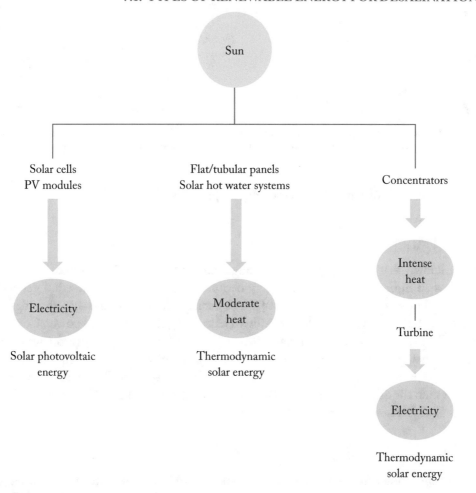

Figure 7.2: Methods of exploiting solar energy (reprinted from [441] with permission).

mountains and islands provide the ample opportunities to use wind energy in desalination. Almost each country in the world has such areas enriched in wind resources. Due to uncertainty in petroleum prices and emission of greenhouse gases, the applications over wind energy expands and grows rapidly [440]. The energy from wind turbines can serve desalination plant by providing electricity or mechanical power. Wind turbines represent a mature technology and have been in operation in many countries across the globe for a long time. Many different types of wind turbines have been developed. Most of these problems, however, are solved by the installation of offshore wind parks. Selection of location for wind operated desalination facilities is crucial. At current stage of development, the site for wind desalination unit must have ample air and saline water resources and scarce water supply.

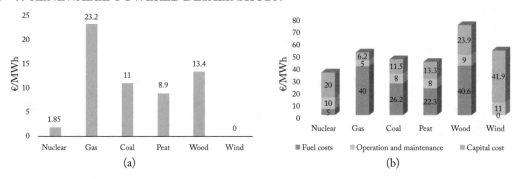

Figure 7.3: Methods of exploiting solar energy (reproduced from [449]).

7.1.4   WAVE ENERGY

It's been known that ocean waves possess huge energy that can be exploited. Waves can be regarded as an indirect form of solar energy because they originate by the wind generated by uneven distribution of solar radiation [440]. High density of seawater (and therefore high energy associated with its movement) adds attraction in harnessing the ocean tidal power. Coastal areas with readily available seawater and wave energy provide the ideal platform for wave energy-driven desalination. Effective utilization of wave energy requires its conversion into other forms of energy such as mechanical or electrical [440, 444]. The conversion efficiencies for wave energy are higher than the other renewable counterparts. In contrast to solar and wind energy, the wave energy is almost absent from the market.

7.1.5   NUCLEAR ENERGY

Nuclear energy comes from splitting atoms in a reactor to heat water into steam, turn a turbine and generate electricity. It currently represents the second largest source of low CO_2 emission energy. Many countries around the globe have nuclear power plants. As of 2016, a total of 441 nuclear reactors were operated in more than 30 countries with a total capacity of 382.9 GW(e) giga-watt (electrical). Among these, 68 reactors are still under construction; 45 of which are in Asia alone, with a total capacity of 67.4 GW(e). Recent studies indicated that global nuclear power capacity will reach 511 GW(e) in 2030, compared to a capacity of around 370 GW(e) in 2009 [445–448].

It is proved that the electricity generated through nuclear power is the lowest-cost electricity supply option in many markets due to the low fuel costs [449]. The analyses reported the different fuels prices and electricity generation costs [449]. As can be clearly seen from Figure 7.3, the electricity generated through nuclear power plants has achieved the lowest cost among the all the alternatives considered. The study also found that the cost of nuclear electricity is insensitive to the changes of nuclear fuel price [450].

The operational mechanism of nuclear energy is the same as that for the traditional fuel-based power station, however, the heat is generated by a chain reaction taking place inside a nuclear reactor. CO_2-emission-free nature of the process puts it in the category of "green technology," although the disposal of radioactive waste and operational safety concerns make it debatable. However, the low cost still makes it attractive. As compared to solar and wind energy, nuclear power ensures consistent and reliable power supply. The current contribution of nuclear power to total global energy demand is $\sim$ 11% [440]. Finite nature of fossil fuels, their environmental concerns and limited technological developments in harnessing the energy from traditional renewable energy resources argue in favor of increasing use of renewable energy in near future in many sectors including desalination [440].

Nuclear desalination appears to be a feasible and a promising option to power desalination plants at reasonable costs. Nuclear desalination can be defined as the production of freshwater from brackish or seawater by using a nuclear reactor as a source of energy for desalination facility. The nuclear facility can be solely dedicated for production of freshwater or a part of energy from the reactor can also be applied for energy production in context of cogeneration plant. The energy can be used in form of thermal or electrical output [440, 450]. A growing interest in using nuclear energy for desalination has been observed worldwide in the past decades. This has been inspired by several parameters and developments including economic competitiveness of nuclear energy, limiting resources of fossil fuels and their environmental concerns, rapidly growing energy demand and significant orientation of industrial research around the nuclear technology [440].

7.1.6   BLUE ENERGY

Osmotic power, salinity gradient power or blue energy is the energy available from mixing two solutions of different salinity [440, 451, 452]. "Salinity energy" stored as the salinity difference between seawater and freshwater is a large-scale renewable resource that can be harvested and converted to electricity, but extracting it efficiently as a form of useful energy remains a challenge. With the development of membrane science and technology, membrane-based techniques for energy extraction from water salinity, such as pressure-retarded osmosis and reverse electro-dialysis, have seen tremendous development in recent years. Meanwhile, many other novel methods for harvesting energy from water mixing processes, such as electrochemical capacitor and nano-fluidic energy harvesting systems, have been proposed. Worldwide, the potential for energy extraction from this "salinity potential" resource (for all river effluents combined) amounts to around 2.4 $\sim$ 2.6 TW [453, 454], close to present-day global electricity consumptions [455].

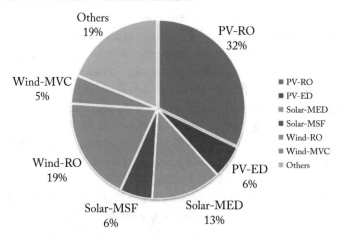

Figure 7.4: Current landscape of renewable energy desalination worldwide (reprinted with permission from [456]).

7.2   RENEWABLE ENERGY POWERED DESALINATION

7.2.1   RENEWABLE ENERGY DRIVEN RO SYSTEMS

Renewable energy is an attractive solution to reduce RO plants carbon footprint, decrease their running costs as well as to eliminate the link between water prices and fuel costs. Desalination by RO is widely considered for renewable energy powered desalination applications due to its low SEC compared to other processes [437, 440, 456]. PV-RO and wind-RO are the most widely deployed technologies for renewable energy powered desalination contributing 32% and 19% of the field, respectively; shown in Figure 7.4. Whereas the power output of solar-PVs and wind turbines is intermittent and fluctuating, commercial RO plants are designed to work at constant flow, pressure and power level. Steady-state operation is considered economical for grid-powered RO plants, as it maximizes production capacity and makes good use of invested capital. In addition, it is easier to maintain the product water quality and manage the membrane fouling. As such, earlier studies of renewable energy-driven RO plant often included a backup, such as a direct connection to an electric grid, an energy storage system, or a Diesel generator to operate the RO plant with constant power [456].

Solar energy is initially converted into electricity to power the pumps while the shaft energy from wind turbine can directly operate high pressure pump as shown in Figure 7.5. PV and thermal solar driven RO units have been installed in various parts globally with capacity varying from less than one m³/day to several hundred cubic meters per day. The specific energy consumption (SEC) of these units ranges from 0.9–29.1 kWh/m^3 and 2.4–17.9 kWh/m^3 for brackish and seawater, respectively [15, 440]. The success of solar-PV as a driver for RO plants is attributed to four factors [457]: (1) the modularity of PV systems offers implemen-

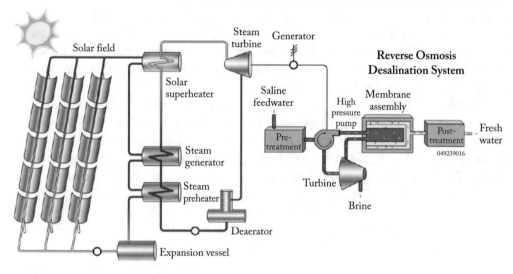

Figure 7.5: Schematic diagram of solar/wind driven RO system (reprinted from [327] with permission).

tation with RO on different scales and their capacity can be increased after initial installation, (2) PVs require low maintenance and offer a long lifetime of 20 years [458], (3) areas that demand high water consumption usually have high solar radiation intensity which makes PVs well matched to the application, and (4) the somewhat predictable bell-shaped diurnal solar irradiance curve, compared to the random variation of wind power, makes it easier to schedule the plant operation during daytime and use water storage instead of energy storage to meet night-time demand [456]. With the decrease in PV costs, PV-RO systems have become more feasible, depending on solar resource availability, RO system demand, water characteristics and local government policies.

A solar powered desalination project based on RO technology has been launched in Dubai recently [440]. The plant produces 30 m^3 of freshwater per day at consumption of 2.8 kWh/m^3 of product water. Relatively high cost of renewable energy (particularly from PV cells) than traditional sources combined with limited operational hours make renewable desalination through RO more expensive [440]. To ensure the uninterrupted and regular power supply around the clock, these plants have been equipped with lead acid batteries that store electricity. However, these batteries have short life, particularly in hot climates, and therefore require regular replacement that increases the overall desalination cost significantly. These batteries also suffer from power losses during charging and discharging cycles and their efficiency goes on decreasing with age. Moreover, the disposal of these batteries causes environmental concerns. Mohamed et al. [459] compared the performance of a RO plant using batteries for energy storage against another plant that is directly coupled to a PV array. The directly coupled plant offered less complexity since there is no need for batteries or a charge controller. Soric et al. [440, 460] studied the possibility of powering PV-RO without using storage batteries with low energy impact. The main component of the prototype is the energy regulator that ensures smooth supply of energy to RO system. Another approach to ensure a full-day operation is to assist the PV system with a conventional energy source such as a Diesel engine [456, 458].

Overall, with sufficient solar resource, PV-RO systems are economically more feasible than Diesel-powered systems. In order to avoid the energy losses related with the transformation of heat energy into electric energy and to eliminate the need of batteries, the concept of Rankin cycle has been introduced [461], where the solar collectors are used to evaporate a low boiling point fluid (generally organic) which drives the shaft of RO pump. Water production cost for the PV-RO system was significantly lower than that of the organic-solar Rankine-RO system at 7.77 €/m^3 compared to 12.53 €/m^3. Caldera et al. suggests that RO unit powered with hybrid PV–wind–battery and power-to-gas power plants are able to fulfill the global water demand of 2030 at levelized water cost of 0.59–2.81 €/m^3 [463].

Low operating cost, high efficiency and energy availability, especially for coastal areas, make wind turbines a successful and clean choice to power RO plants [456, 464–466]. The wind energy lowers both carbon footprint and water production costs. Wind-driven RO plant can be competitive with thermal desalination in United Arab Emirates [467]. Similar to solar energy, different options have been suggested to provide smooth energy output and water supply from wind energy driven units. Combination of wind generator with photovoltaic energy has been proposed to supply the uninterrupted power [468]. However, further deployment of wind turbines requires certain challenges to be addressed. Wind turbines have to gain social acceptance and improved public perception due to their aerodynamic noise and visual impact [456]. Wind turbines can be used with solar-PV to extend energy availability to include night time and overcast days, providing more consistent output [456].

A 2.2 kW small wind turbine has been used to drive RO desalination unit [469]. A computer model was developed in order to find optimum wind turbine power that yields maximum freshwater production rate. The model has been verified and supported by an experimental setup. The results showed that increasing wind speed up to 11 m/s resulted in higher water production rate, after which the rate dropped slightly as wind speed increased. Moreover, the results showed that the specific energy consumption of RO desalination system decreased sharply as wind speed increased up to 6 m/s, within wind speed range of 6–11 m/s, the specific energy consumption increased as wind speed increased and beyond 11 m/s it decreased slightly. The results are shown in Table 7.1.

Wave/tidal energy, another possible option for powering RO plants, is generally available where seawater is desalinated. But the harnessing of wave energy is, as with other forms of renewable energy, expensive in terms of capital plant and the effort needed to develop the technology. It may require the intervention of governments or international bodies. Coastal sites with abundant wave energy provide the opportunity to exploit direct pressurization of seawater

Table 7.1: Variation of wind speed with fresh water production and RO desalination specific energy consumption rate [470]

Wind Speed (m/s)	Rate (m³/d)	SEC (kWh/m³)
5	2	4.4
6	5.7	3.15
7	7.5	3.25
8	9.3	3.32
9	10	3.4
10	10.2	3.45
11	10.25	3.4
12	9.8	3.38
13	9.6	3.36
14	9.5	3.38
15	9.4	3.36

for production of freshwater. This principle has been utilized in designing some desalination setups where the converter absorbs wave energy and transfers it to the RO unit [440]. Applications of wind energy in desalination started in the early 1980s. It has been demonstrated that wave powered desalination can reduced the water short fall by 16%, 64%, and 100% in Morocco, Oman and Somalia, respectively [471]. In general, compared to other resources of renewable energy, relatively few investigations have been devoted to practically demonstrate the potential of wave power in desalination. Sharmila et al. [472] proposed a self-sustaining wave energy desalination plant based on Oscillating Water Column principle, based on alternating compressing and depressurizing the column of air entrapped above the water column by using the wave action. The compressed air is then used to run a turbine and extracted power has been reported to run a RO plant with capacity of 10 m³/day. Folley et al. [473] studied the potential for an autonomous wave-powered desalination by RO plant using a pressure exchanger-intensifier for energy recovery. It was found that a SEC of less than 2 kWh/m³ of distillate can be obtained by using this system over a wide range of seawater conditions, making it technically and economically suitable for seawater desalination. SEC drops down to 1.85 kWh/m³ when recovery rate is maintained at 25–35% due to elimination of pre-treatment requirement and reduced scaling problems.

For nuclear power-driven desalination, RO as standalone process or hybrid combinations are getting more popularity than thermal plants. For RO, the costs have been identified one-third of that of gas driven systems. The feasibility reports based on 150 reactor-years of experience with nuclear desalination shows the product cost in range of $0.7–0.9/m³ [440]. The world's largest

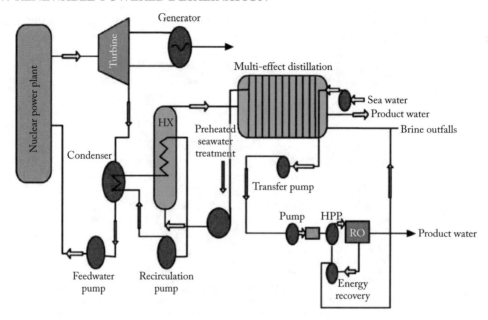

Figure 7.6: A schematic diagram for a nuclear desalination process, HX: Heat Exchanger and HPP: High Pressure Pump (reprinted from [450] with permission).

nuclear desalination plant (hybrid MSF-RO) has been working in Madras Atomic Power Station, southeast India. The plant initially produced most of freshwater through MSF but later on a barge-mounted RO unit was added to improve the overall capacity of the plant. Similarly, a nuclear energy drive RO unit with 450 m³/day capacity is in operation in Pakistan since 2014. However, the major experience comes from installation from Russia, Eastern Europe and Canada [440]. Table 7.2 shows the capacity of some desalination plants at different locations around the world including Japan, Kazakhstan, and India with electrical power capacity exceeding 1000 MW as well as the method of desalination used. It can be clearly seen that PWR reactors are commonly used with RO [450].

The schematic for a desalination process coupled with a nuclear power plant is shown in Figure 7.6 [474]. It is an on-site nuclear-desalination system. The purpose is to generate electricity (to power RO) as well as utilizing the waste heat to produce steam that will be fed into the MED unit. In order to design a nuclear desalination process, the following steps should be performed: (1) proper modeling for the reactor-desalination systems, (2) careful evaluation for the nuclear plant safety, and (3) technical outcomes from the desalination process itself [450].

Numerous studies considered the design of wind-RO plants and presented different approaches to accommodate the variable nature of wind power, as summarized in Table 7.3. Table 7.4 presents a summary of previous studies discussing solar-PV powered SWRO plants. And

Table 7.2: Global nuclear desalination capacities [450]

Plant Name	Location	Gross Power [MW(e)]	Capacity [m³/d]	Energy/ Desalination
Shevchenko	Aktau, Kazakhstan	150	80,000–145,000	LMFR/MED, MSF
Ikata-1,2	Ehime, Japan	566	2000	PWR/MSF
Ikata-3	Ehime, Japan	890	2000	PWR/RO
Ohi-1,2	Fukui, Japan	2 × 1175	3900	PWR/MSF
Ohi-3,4	Fukui, Japan	1 × 1180	2600	PWR/RO
Genkai-4	Fukuoka, Japan	1180	1000	PWR/RO
Genkai-3,4	Fukuoka, Japan	2 × 1180	1000	PWR/MED
Takahama-3,4	Fukui, Japan	2 × 870	1000	PWR/RO
NDDP	Kalpakkam, India	170	6300	PHWR/Hyb. MSF-RO
LTE	Trombay, India	40 [MW(t)]	30	PHWR/LTE (Low temperature evaporation desalination)
Diablo Canyon	San Luis Obispo, USA	2 × 1100	2180	PWR/RO

Note: pressurized water reactor (PWR); pressurized heavy water reactor (PHWR)

Table 7.5 summarizes previous studies of hybrid RE-powered SWRO plants, giving an overview of the current status and trends [456].

Table 7.3 Summary of previous literature discussing wind-RO desalination [456].

7.2.2   RENEWABLE ENERGY POWERED ED SYSTEMS

Sustainable water desalination by ED requires a renewable energy input. A number of studies have been carried out on testing and analysis of an ED system driven by photovoltaics. The performance of a small scale ED in four hydraulic stages and two electrical stages has been tested [492], driven with solar energy converted into DC by photovoltaics having a peak power of 33 W and a nominal voltage of 12 V. It was observed that the quality of the product water was substantially improved at low product flow rates (about 150 gallons/day) with salt rejection reaching up to 99% for the tested NaCl solutions [492].

Ortiz et al. demonstrated the feasibility of desalination of brackish water by means of an ED system directly powered with PV solar panels without the use of batteries [493]. When using 8 PV panels, the desalination process took 0.35 h and the drinkable water production

Table 7.3: Summary of previous literature discussing wind-RO desalination [456]

Ref.	Year	Location	Study	Energy Storage	Wind Turbine Power (kW)	Production Capacity (m³/day)	S.E.C (kWh/m³)	Feed Salinity (mg/L)	Production Cost	Energy Recovery	Plant Description
[469]	2002	Red Sea	Exp.	None	2.2	8.5	3.4	40,000	0.8–3 (£/m³)	Clark pump	—
[475]	2003	Canary Islands	Exp.	Battery and UPS for control system	2 × 230	25 × 8	6.9	Seawater	—	None	Eight identical plants connected in parallel. Each plant has three pressure vessels with three RO membranes each.
[476]	2004	Colombia	Exp.	None	1.5	0.4	—	35,000	—	None	One RO membrane.
[477]	2009	Not based on wind data	Theo.	None	Power from main: 12.5	30.5	3.2–4.22	35,646	—	Presented by an 85% efficiency	Four RO membranes in series.
[478]	2011	Canary Islands	Exp.	Flywheel-batteries	225	1000	2.7	38,170	—	RO Kinetic®	Three RO trains. Each train has 2, 6, and 10 pressure vessels, respectively. Each pressure vessel includes seven RO membranes in series.
[479]	2012	Cape Verde	Theo.	None	275	200–600	4.3	Seawater	1.09 (€/m³)	Hydraulic turbocharger	Two RO trains.
[480]	2015	Canary Islands	Theo	None	15	5.2–19.4	10–14.5 11.3–16.9	35,200 39,800	—	None	Two pressure vessels connected in parallel. Three modules per PV.
[464]	2015	Turkey	Theo.	None	30	24	4.38	43,528	2.96–6.46 ($/m³)	None	Six RO membranes in series.
[481]	2015	Norway	Exp.	None	5	7.5	4.24	35,000	—	None	Eight RO membranes in series.
[482]	2015	Canary Islands	Exp.	None	Power from main 5.5–21.5	45.6–120	4–5.5	32,237 ppm	—	None	One pressure vessel with six RO membranes connected in series.

Table 7.4: Summary of previous literature discussing PV-RO desalination [456]

Ref.	Year	Location	Study	Energy Storage	PV Power (kW)	Production Capacity (m³/day)	S.E.C (kWh/m³)	Feed Salinity (mg/L)	Production Cost	Energy Recovery	Plant Description
[462]	2007	Athens	Exp.	No	0.85	0.35	4.6	32,738	7.8 €/m³	Clark pump	Two RO membranes in series.
[458]	2008	United Arab Emirates	Theo.	No	17.9	20	7.33	45,000	7.34 $/m³	Yes	Two-stage system with booster pump between stages. Two RO membranes in series per stage.
[462]	2008	Thirasia Island, Greece	Exp.	No	0.846	2.4	3.8-6	22,000	7.77 €/m³	Clark pump	Two RO membranes in series.
[483]	2011	USA	Exp.	Batteries to power the control electronics	0.23	0.3	4-2.5	35,000	4.7-6.62 $/m³	Clark pump	One RO membrane.
[460]	2012	Marseille, South of France	Exp.	No	0.5	0.75-1.02	—	25,000	—	Clark pump	Two RO membranes in series.
[484]	2012	Australia	Theo.	Compare with/ without batteries	0.7	With battery: 0.054 Without battery: 0.047	—	Seawater	—	None	One RO membrane (Commercial unit).
[485]	2013	USA	Exp.	No	0.23	0.3-0.45	—	Seawater	—	Dual-piston pressure exchanger	One RO membrane.
[486]	2015	India	Theo.	No	0.667	0.7	—	35,000	—	None	One RO membrane.
[487]	2016	Greece Spain UAE	Exp.	No	10-20	Single unit 12-16.8 3 identical units	5.2-5.8	37,500	—	Axial piston motor	Three identical units connected in parallel. Four RO membranes in series each.

Table 7.5: Summary of previous literature discussing hybrid Wind-PV RO desalination [456]

Ref.	Year	Location	Study	Backup System	Hybrid System	Production Capacity (m³/d)	S.E.C (kWh/m³)	Feed Salinity (mg/L)	Production Cost	Energy Recovery	Plant Description
[488]	2001	Israel	Exp.	Battery + DE	WT PV	3	—	3500–5000	—	None	Two RO membranes in series
[489]	2002	Libya	Theo.	Grid-connected	WT PV Grid	300	5.6	Seawater	2.3 €/m³	None	Two RO trains
[490]	2010	Greece	Theo.	Pumped storage	WT PV Hydro.	3840 Based on hourly average	3	Seawater	2.53 €/m³	—	—
[491]	2013	Saudi Arabia	Theo.	Battery	WT PV	5	8–20	—	3.693–3.812 $/m³	—	—

capacity of the system reached approximately 0.29 m³/h. However, when the number of PV panels was reduced by half, i.e., to 4 PV panels, the time of desalination process was increased to 0.45 h whereas the drinkable water production capacity was reduced to 0.22 m³/h. In an ED plant coupled with a PV panel of 450 W peak power, operated at 80 V, the plant equipped with 42 cell pairs (dimensions of 30×60 cm) reached a capacity of about 1000 L/day at product water salinity below 1000 ppm [440, 494]. Regarding energy consumption, it was found that the use of PV-ED lies in the range of 0.02–0.03 kg CO^2/m³ (only due to energy supply) for a system operating with a brackish water having a concentration of 2500–5000 mg/L and energy consumption of 0.49–0.91 kWh/m³ [495]. This is one order of magnitude less than that of the grid mix supply.

The largest solar powered ED plant is located in Fukue City, Japan which was installed in 1990. The plant, consisting of a 65 kW$_p$ PV array with 1.2 Ah of storage, has an average water production capacity of 200 m³/d with an energy consumption of 0.6–1 kWh/m³ [496]. Wright et al. designed a village-scale ED system that runs on solar energy (PV-ED). The PV-ED system was justified to be cost effective and energy efficient means of desalinating groundwater in rural India [497]. It was demonstrated that the ED system requires 75% less specific energy than RO at 1000 mg/L and 30% less than RO at 3000 mg/L. When using PV-powered system, the SEC of ED reaches 50% lower than RO system at 2000 mg/L, implying the feasibility of potential expansion of PV-ED as a desalination system in those areas. From the data of Table 7.6, it can be concluded that despite of the wide range of plant capacities (1–200 m³/day), most of the installations were of small size and desalinated brackish water all over the world. The SEC is between 0.4 and 4 kWh/m³ and most of the plants used PV alone, without any storage. The economic feasibility of PV-ED depends upon capital and operational costs which, in turn, depends

Table 7.6: Renewable energy driven ED desalination plants installed worldwide [495]

Location	National Subsidies	Capacity (m³/d)	Year	Feed TDS (mg/L)	Cell Pairs (number)	Water Cost (€/m³)	SEC (Kwh/m³)	PV System (kWp)
Spencer Valley, Mexico	Yes	2.8	1986	1,000	–	11.9	0.82	1 (tracking) + 2.3 (stationary)
Thar Desert, India	Yes	1	1986	5,000	42	–	1	0.45
Ohsima Island, Japan	Yes	10	1986		250	4.3	–	25
Fukue, Japan	Yes	200	1990	700	–	–	0.6–1	65+ Batteries
New Mexico, Mexico	Yes	18	1996	900	–	–	0.8	2.3+ Batteries
Isa, Bahrain		1.14	2002	3,300	24	–	–	0.132
Alicante, Spain	Yes	1.32	2006	2,000	80	–	–	0.272
Alicante, Spain	Yes	13.7	2008	4,473	–	0.14–0.32	1.33–1.47	38.45
La Luz, Mexico	Yes	15	1979	–	–	–	–	5
Colorado, USA	Yes	2.8	1995	–	–	–	–	2.3
Canary Islands, Spain	Yes	4	2013	2,240–3,392	340	–	0.618	3.69
Zaragoza, Spain	Yes	4.3	2013	580–10,000	10	–	0.4–0.6	0.103–0.308

upon operational efficiency. The capital cost is mainly determined by the required membrane area (RO module or ED stack), pumps, piping, valves, storage tanks, electrical instrumentation and control equipment, energy recovery devices and water pre-treatment equipment. The operating cost is dependent on the energy consumption and the cost of the membrane and pre-filter replacement, pre-treatment chemicals and general maintenance. Sustainability issues related to the Photovoltaic-ED energy for the desalination of brackish water have been assessed considering environmental, economic and social issues [495]. Matching of the intermittent output of renewable energies with water demand, a lifetime of membranes, efficiency of solar panels and the high production cost of freshwater are the main barriers preventing PV-ED to market penetration [440].

7.2.3   RENEWABLE ENERGY DRIVEN MD SYSTEMS

A low-grade heat source is needed to supply energy for a MD process. In this way, renewable energy sources are an attractive solution, because they have lower environmental impacts and lower greenhouse gas emissions compared to fossil fuels [409]. Solar-powered membrane distillation (SP-MD) processes, which use the solar radiation to supply energy to drive the MD system, are being widely investigated and implemented by many researchers and practitioners [498]. Typically, the solar energy required to provide the thermal energy to the MD process can be achieved using collector technologies such as flat plate collectors (FPCs), evacuated tube collectors (ETCs), compound parabolic concentrators (CPCs), salt-gradient solar ponds (SGSPs) or solar stills [499]. Besides the thermal energy, MD requires electricity to drive pumps and other electric devices. Electricity could be supplied from the electric grid or using an auxiliary diesel generator—typically referred to as the assisted way, or with PV collectors with direct current battery cells and electric current inverters, also known as the stand-alone way [500]. The technology can be of particular interest for the regions with abundant resources of solar energy but less availability of fresh water such as Middle East, Australia, Spain, Italy, Greece, South Asia, and some parts of North America [440]. A scheme of a solar-powered MD system is shown in Figure 7.7.

There are two ways to connect solar collectors with MD modules: the single-loop system and the two-loop system [501]. In the single-loop system (or compact layout), the solar collector is directly connected to the membrane module (Figure 7.8a). In this system, special care must be taken to avoid corrosion problems because salty water is recirculated through the single loop [501]. However, it is suitable for small systems that have low thermal energy demand because the feed flow rate and the temperature can be changed quickly without creating relevant transient effects in the desalination module [501]. In the two-loop system (Figure 7.8b), the solar collector and the MD module are connected by a heat exchanger, and also the system could have heat storage, which allows extending the operation time of the MD system for some hours beyond the sunset [501].

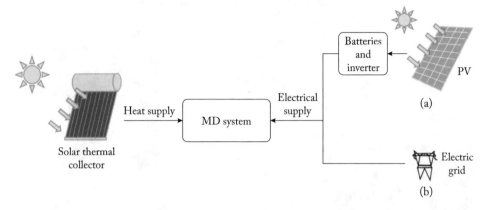

Figure 7.7: Schematic of a solar-powered MD system for a stand-alone way (a) and for an assisted way (b) (reprinted from [409] with permission).

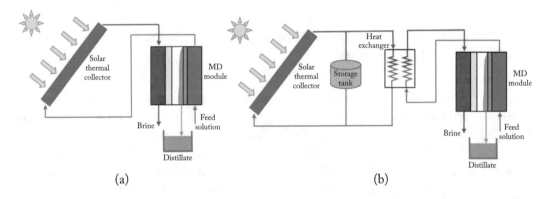

Figure 7.8: Scheme of a solar-powered MD system: (a) single-loop system and (b) two-loop system (reprinted from [409] with permission).

Banat et al. [502] have analyzed the technical feasibility of an MD system integrated with a solar still for production of freshwater. The designing and manufacturing of a small scale autonomous solar driven air gap MD unit for desalination in arid and remote regions have also been investigated. Solar thermal collectors have been used for heating the feed solution while electricity is supplied by solar PV. A very good quality of distillate has been claimed at expense of 200–300 kWh/m^3 of SEC. The cost was a strong function of membrane and plant life and was equal to \$15/m^3 [440].

While most of the systems used for solar driven MD are based on air gap MD, vacuum membrane distillation (VMD) systems can offer the advantage of high flux even at low feed temperatures. Solar heated hollow fiber VMD systems in Hangzhou region of China have been

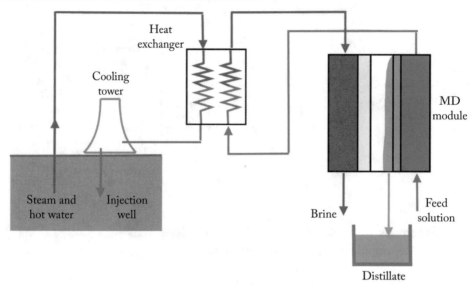

Figure 7.9: A simplified scheme of a geothermal power plant coupled with an MD system (reprinted from [409] with permission).

tested [503]. Mericq et al. [504] analyzed the use of salinity gradient solar ponds and solar collectors for MD for two configurations: heating of feed before entering into the modules and submerging the module inside salinity gradient solar ponds or insertion of solar collector inside the module. The application of solar collector was pointed out as the most promising option with water flux as high as 142 LMH. Chen et al. [505] introduced the design of a direct contact MD module equipped with solar absorber. The absorber directly uses solar radiation to heat the feed within the module. It was suggested the use of high incident solar radiation, thermosiphon and thin spacers to further improve the performance of the module. Table 7.7 summarizes some studies since 2010 that couple MD with solar energy. Details can be referred to [409].

Uninterrupted supply of heat makes geothermal energy particularly attractive for desalination applications. The intensity of heat from geothermal sources is generally not suitable for power production that further adds the motivation of using it for desalination objectives. Also, many regions with abundant availability of geothermal energy are facing the problem of water scarcity. Conventional thermal processes such as MSF require very high temperature for their operation which is generally not achievable by exploiting geothermal energy. RO is not suitable candidate for geothermal driven desalination due to low grade of geothermal heat and due to necessity to convert heat input into electric input that renders a lot of energy losses. This leaves MD as the most suitable technology to exploit geothermal energy for desalination [440]. As an example, a simplified scheme of a MD system driven by geothermal energy is shown in Figure 7.9. SEC indicated from the literature varies from 1 to ~ 9000 kWh/m³ [498], clearly indi-

Table 7.7: Solar energy driven MD systems [409]

MD Configuration	Energy Source	Study	Feed Solution	Location	Ref.
PGMD	PV and FPC	Pilot plant	Seawater (from the Red Sea)	Aqaba, Jordan	[506]
DCMD	SGSP	Theoretical	NaCL solution (50 and 75 g/L)	Nevada, USA (Walker Lake)	[507]
VMD	SGSP and SC	Theoretical	Seawater	Gabès, Tunisia	[508]
AGMD, PGMD, VMD and multi-stage MD	ETC	Experimental and theoretical	Seawater (35 g/l)	Palermo, Italy (Università di Palermo)	[509]
MEMD	HCPVT	Experimental and theoretical	Saline and brackish water	Zurich, Switzerland (IBM Zurich research Laboratory)	[510]
DCMD, AGMD and VMD	PV and FPC	Theoretical	Seawater	Arabian Gulf countries	[500]
AGMD and multi-stage AGMD	CPC	Pilot plant	Seawater (1 and 35 g/l)	Almería, Spain (Plataforma Solar de Almería, PSA)	[511]
AGMD	Solar absorber is simulated with a thermostat	Experimental and theoretical	Deionized water	N/A	[512]
PGMD	FPC and PV	Pilot plant	Seawater	Gran Canaria Island, Spain (Instituto Tecnológico de Canarias, ITC, in Playa de Pozo Izquierdo).	[513]
V-MEMD	SC and PV/ diesel	Experimental	Seawater	N/A	[514]
PGMD	ETC (heat pipe type)	Experimental and theoretical	Seawater	N/A	[515]
AGMD and PGMD	FPC	Pilot plant	Ground water (28000 ppm)	Amarika, Namibia (Etosha basin, remote inland location)	[516]
			Sea water Atlantic ocean (35000 ppm)	Gran Canaria Island, Spain (Instituto Tecnológico de Canarias, ITC, in Playa de Pozo Izquierdo).	[516]
V-MEMD	PV and ETC	Pilot plant	Seawater and brackish water	Saudi Arabia	[517]
VMD	ETC	Experimental	Seawater	Xiamen, China	[518]
AGMD, LGMD, V-MEMD	CPC / FPC	Experimental	Seawater and brackish water	Almería, Spain (Plataforma Solar de Almería, PSA)	[519]
DCMD	SC	Experimental and theoretical	Seawater	N/A	[520]
DCMD	SGSP	Experimental and theoretical	Saline water (13 g/l)	RMIT Bundoora east campus, Australia	[501]
DCMD	SGSP	Experimental	Fresh water	N/A	[521]
LGMD	CPC-reactor	Experimental	Urban wastewater with biological contaminants	Almeria, Spain (Plant at El Bobar)	[522]
DCMD	FPC	Pilot plant	Seawater	Jeddah, KSA	[523]
AGMD	FPC	Experimental and theoretical	NaCL solution (3.5 wt-%)	N/A	[524]

Note: CPC: compound parabolic concentrator; ETC: evacuated tube collector; FPC: flat-plate collector; HCPVT: high concentration photovoltaic thermal; PV: solar photovoltaic; SC: solar collector (not specified); SGSP: salt-gradient solar pond.

cating the underlying ambiguities. The observed dispersion can be associated with several factors including the use of different solutions (seawater, brine, etc.), different MD configurations, different energy sources, non-optimized operating conditions, different membrane materials and performance, different plant size, capacity and life, lack of information about pre-treatment requirements, etc. These factors clearly indicate the future direction of research to determine the precise specific energy demands of the process [440].

Bibliography

[1] R. Goldstein and W. Smith. Water and sustainability (volume 4): U.S. electricity consumption for water supply and treatment—the next half century. *Electric Power Research Institute*, 2002. 1, 52, 71

[2] J. C. Crittenden, R. R. Trussell, D. W. Hand, K. J. Howe, and G. Tchobanoglous. *MWH's Water Treatment: Principles and Design*, John Wiley & Sons, 2012. DOI: 10.1002/9781118131473. 3, 4, 6, 7, 71

[3] J. Mulder. *Basic Principles of Membrane Technology*, Springer Science & Business Media, 2012. DOI: 10.1007/978-94-017-0835-7. 3, 4, 5, 6, 7

[4] E. M. V. Hoek, J. Allred, T. Knoell, and B.-H. Jeong. Modeling the effects of fouling on full-scale reverse osmosis processes. *J. Memb. Sci.*, 314:33–49, 2008. DOI: 10.1016/j.memsci.2008.01.025. 5, 6, 10, 12, 15, 31

[5] Y. A. Le Gouellec and M. Elimelech. Calcium sulfate (gypsum) scaling in nanofiltration of agricultural drainage water. *J. Memb. Sci.*, 205:279–291, 2002. DOI: 10.1016/S0376-7388(02)00128-X. 5, 6

[6] L. Song, J. Y. Hu, S. L. Ong, W. J. Ng, M. Elimelech, and M. Wilf. Emergence of thermodynamic restriction and its implications for full-scale reverse osmosis processes. *Desalination*, 155:213–228, 2003. DOI: 10.1016/s0011-9164(03)00300-x. 7

[7] G. Guillen and E. M. V. Hoek. Modeling the impacts of feed spacer geometry on reverse osmosis and nanofiltration processes. *Chem. Eng. J.*, 149:221–231, 2009. DOI: 10.1016/j.cej.2008.10.030. 6, 7, 24

[8] E. M. V. Hoek and M. Elimelech. Cake-enhanced concentration polarization: A new fouling mechanism for salt-rejecting membranes. *Environ. Sci. Technol.*, 37:5581–5588, 2003. DOI: 10.1021/es0262636. 6, 7, 10, 11

[9] S. Bhattacharjee. Concentration polarization: Early theories, (n.d.). http://appliedchem. unideb.hu/UnitOperation3/Membraneseparation/ConcentrationPolarization_FINAL_ 7-11-17.pdf 7

[10] A. Jawor and E. M. V. Hoek. Effects of feed water temperature on inorganic fouling of brackish water RO membranes. *Desalination*, 235:44–57, 2009. DOI: 10.1016/j.desal.2008.07.004. 7

[11] E. M. V. Hoek. Reverse osmosis membrane biofouling: Causes, consequences and countermeasures, (n.d.). www.waterplanet.com/wp-content/.../WPI_ROBiofouling_WhitePaper_7-13-17.pdf 6, 7, 11, 73

[12] S. Kim and E. M. V. Hoek. Modeling concentration polarization in reverse osmosis processes. *Desalination*, 186:111–128, 2005. DOI: 10.1016/j.desal.2005.05.017.

[13] E. M. V. Hoek, A. S. Kim, and M. Elimelech. Influence of crossflow membrane filter geometry and shear rate on colloidal fouling in reverse osmosis and nanofiltration separations. *Environ. Eng. Sci.*, 19:357–372, 2002. DOI: 10.1089/109287502320963364. 7, 10, 49, 52, 53, 57

[14] G. B. Van den berg and C. A. Smolders. Flux decline in membrane processes. *Desalination*, 25:101–133, 1988. DOI: 10.1016/0011-9164(90)85023-4. 7

[15] L. F. Greenlee, D. F. Lawler, B. D. Freeman, B. Marrot, and P. Moulin. Reverse osmosis desalination: Water sources, technology, and today's challenges. *Water Res.*, 43:2317–2348, 2009. DOI: 10.1016/j.watres.2009.03.010. 8, 9, 12, 125, 130

[16] L. Henthorne and B. Boysen. State-of-the-art of reverse osmosis desalination pre-treatment, 356:129–139, 2015. DOI: 10.1016/j.desal.2014.10.039. 9

[17] W. B. Russel. Colloidal dispersions, 1989. DOI: 10.1017/cbo9780511608810. 9, 92

[18] J. J. Morgan and W. Stumm. *Aquatic Chemistry*, Wiley, 1970. DOI: 10.2307/2260132. 9

[19] V. L. Snoeyink and D. Jenkins. *Water Chemistry*, John Wiley, 1980. 9

[20] A. Alhadidi, A. J. B. Kemperman, R. Schurer, J. C. Schippers, M. Wessling, and W. G. J. Van der Meer. Using SDI, SDI+ and MFI to evaluate fouling in a UF/RO desalination pilot plant. *Desalination*, 285:153–162, 2012. DOI: 10.1016/j.desal.2011.09.049. 9

[21] T. M. C. Fritzmann, J. Löwenberg, and T. Wintgens. State-of-the-art of reverse osmosis desalination pre-treatment. *Desalination*, 356:129–139, 2007. DOI: 10.1016/j.desal.2006.12.009. 72

[22] A. Alhadidi, A. J. B. Kemperman, J. C. Schippers, B. Blankert, M. Wessling, and W. G. J. van der Meer. SDI normalization and alternatives. *Desalination*, 279:390–403, 2011. DOI: 10.1016/j.desal.2011.06.042.

[23] A. Alhadidi, A. J. B. Kemperman, J. C. Schippers, M. Wessling, and W. G. J. van der Meer. SDI: Is it a reliable fouling index? *Desalin. Water Treat.*, 42:43–48, 2012. DOI: 10.1080/19443994.2012.683102. 9

[24] J. A. Brant and A. E. Childress, Assessing short-range membrane—colloid interactions using surface energetics, 203:257–273, 2002. DOI: 10.1016/s0376-7388(02)00014-5. 10

[25] M. R. Wiesner and S. Chellam. Peer reviewed: The promise of membrane technology. *Environ. Sci. Technol.*, 33:360A–366A, 1999. DOI: 10.1021/es993008l. 10, 71

[26] R. D. Cohen and R. F. Probstein. Colloidal fouling of reverse osmosis membranes, 1986. DOI: 10.1016/0021-9797(86)90252-3. 10

[27] G. D. Bixler and B. Bhushan. Biofouling: Lessons from nature. *Philos. Transactions on R. Soc. A. Math. Phys. Eng. Sci.*, 370:2381–2417, 2012. DOI: 10.1098/rsta.2011.0502. 11, 49

[28] T. Mattila-Sandholm and G. Wirtanen. Biofilm formation in the industry: A review. *Food Rev. Int.*, 8:573–603, 1992. DOI: 10.1080/87559129209540953. 11, 71

[29] B. Meyer. Approaches to prevention, removal and killing of biofilms. *Int. Biodeterior. Biodegradation.*, 51:249–253, 2003. DOI: 10.1016/s0964-8305(03)00047-7. 11

[30] M. Herzberg, S. Kang, and M. Elimelech. Role of extracellular polymeric substances (EPS) in biofouling of reverse osmosis membranes. *Environ. Sci. Technol.*, 43:4393–4398, 2009. DOI: 10.1021/es900087j. 11

[31] M. Herzberg and M. Elimelech. Biofouling of reverse osmosis membranes: Role of biofilm enhanced osmotic pressure. *J. Memb. Sci.*, 295:11–20, 2007. DOI: 10.1016/j.memsci.2007.02.024. 11

[32] K. Majamaa, J. E. Johnson, and U. Bertheas. Three steps to control biofouling in reverse osmosis systems. *Desalin. Water Treat.*, 42:107–116, 2012. DOI: 10.5004/dwt.2012.2466. 11

[33] C. Tzotzi, T. Pahiadaki, S. G. Yiantsios, A. J. Karabelas, and N. Andritsos. A study of $CaCO_3$ scale formation and inhibition in RO and NF membrane processes. *J. Memb. Sci.*, 296:171–184, 2007. DOI: 10.1016/j.memsci.2007.03.031. 12

[34] D. L. Shaffer, M. E. Tousley, and M. Elimelech. Influence of polyamide membrane surface chemistry on gypsum scaling behavior. *J. Memb. Sci.*, 525:249–256, 2017. DOI: 10.1016/j.memsci.2016.11.003. 12

[35] A. Antony, J. H. Low, S. Gray, A. E. Childress, P. Le-Clech, and G. Leslie. Scale formation and control in high pressure membrane water treatment systems: A review. *J. Memb. Sci.*, 383:1–16, 2011. DOI: 10.1016/j.memsci.2011.08.054. 12

[36] R. J. Xie, M. J. Gomez, Y. J. Xing, and P. S. Klose. Fouling assessment in a municipal water reclamation reverse osmosis system as related to concentration factor. *J. Environ. Eng. Sci.*, 3:61–72, 2004. DOI: 10.1139/s03-054. 13, 15

[37] A. Matin, F. Rahman, H. Z. Shafi, and S. M. Zubair. Scaling of reverse osmosis membranes used in water desalination: Phenomena, impact, and control; future directions. *Desalination*, 455:135–157, 2019. DOI: 10.1016/j.desal.2018.12.009. 12, 13

[38] J. Benecke, M. Haas, F. Baur, and M. Ernst. Investigating the development and reproducibility of heterogeneous gypsum scaling on reverse osmosis membranes using real-time membrane surface imaging. *Desalination*, 428:161–171, 2018. DOI: 10.1016/j.desal.2017.11.025. 12

[39] T. Tong, A. F. Wallace, S. Zhao, and Z. Wang. Mineral scaling in membrane desalination: Mechanisms, mitigation strategies, and feasibility of scaling-resistant membranes, 579:52–69, 2019. DOI: 10.1016/j.memsci.2019.02.049. 12, 15

[40] E. E. A. Ghafour. Enhancing RO system performance utilizing antiscalants. *Desalination*, 153:149–153, 2003. DOI: 10.1016/s0011-9164(02)01117-7. 12

[41] R. J. Petersen. Composite reverse osmosis and nanofiltration membranes. *J. Memb. Sci.*, 83:81–150, 1993. DOI: 10.1016/0376-7388(93)80014-o. 15

[42] K. P. Lee, T. C. Arnot, and D. Mattia. A review of reverse osmosis membrane materials for desalination—development to date and future potential. *J. Memb. Sci.*, 370:1–22, 2011. DOI: 10.1016/j.memsci.2010.12.036. 15

[43] K. Hayashi. Interfacial polymerization. *J. Synth. Org. Chem. Japan.*, 18:206–214, 2011. DOI: 10.5059/yukigoseikyokaishi.18.206. 15

[44] Lenntech, Fluid Systems ®Roga ®-HR 4"Elements Cellulose Acetate RO Elements Product Description, 2009.

[45] Pentair, HFW1000 membrane modules | X-Flow, (n.d.). https://xflow.pentair.com/en/products/hfw1000

[46] TriSep X-20, (n.d.). https://www.lenntech.com/Data-sheets/Trisep-X-20-Lit-L.pdf

[47] Duraslick * Series, (n.d.). https://www.lenntech.com/Data-sheets/GE-Osmonics-Duraslickseries-Low-Fouling-NF-RO-Elements.pdf

[48] Product Data Sheet FILMTEC TM Membranes NF270, (n.d.). https://www.dupont.com/content/dam/Dupont2.0/Products/water/literature/609-00519.pdf

[49] Dow Water Solutions, FILMTECTM Membranes FT30 Reverse Osmosis Membrane Specifications, (n.d.). http://msdssearch.dow.com/PublishedLiteratureDOWCOM/dh_0074/0901b80380074549.pdf?filepath=liquidseps/pdfs/noreg/609-01020.pdf&fromPage=GetDoc

[50] ESPA1, (n.d.) 92058. http://membranes.com/solutions/products/ro/espa/#ESPA1

[51] LG BW 400 ES, (n.d.). https://www.lenntech.com/Data-sheets/LG-BW-400-ES.pdf

[52] D. Li and H. Wang. Recent developments in reverse osmosis desalination membranes. *J. Mater. Chem.*, 20:4551–4566, 2010. DOI: 10.1039/b924553g. 15, 20

[53] S. Beverly, S. Seal, and S. Hong. Identification of surface chemical functional groups correlated to failure of reverse osmosis polymeric membranes. *J. Vac. Sci. Technol. A Vacuum, Surfaces, Film.*, 2002. DOI: 10.1116/1.582308. 15

[54] A. Subramani and E. M. V. Hoek. Biofilm formation, cleaning, re-formation on polyamide composite membranes. *Desalination*, 257:73–79, 2010. DOI: 10.1016/j.desal.2010.03.003. 15

[55] M. Shannon, P. W. Bohn, M. Elimelech, J. G. Georgiadis, B. J. Mariñas, and A. M. Mayes. Science and technology for water purification in the coming decades. *Nature*, 452:301–310, 2008. DOI: 10.1038/nature06599. 15, 19

[56] A. E. Contreras, Z. Steiner, J. Miao, R. Kasher, and Q. Li. Studying the role of common membrane surface functionalities on adsorption and cleaning of organic foulants using QCM-D. *Environ. Sci. Technol.*, 45:6309–6315, 2011. DOI: 10.1021/es200570t. 15

[57] G. Kang, M. Liu, B. Lin, Y. Cao, and Q. Yuan. A novel method of surface modification on thinfilm composite reverse osmosis membrane by grafting poly(ethylene glycol). *Polymer (Guildf)*, 2007. DOI: 10.1016/j.polymer.2006.12.046. 15

[58] G. R. Guillen, Y. Pan, M. Li, and E. M. V. Hoek. Preparation and characterization of membranes formed by nonsolvent induced phase separation: A review. *Ind. Eng. Chem. Res.*, 2011. DOI: 10.1021/ie101928r. 15

[59] M. M. Pendergast and E. M. V. Hoek. A review of water treatment membrane nanotechnologies. *Energy Environ. Sci.*, 4:1946–1971, 2011. DOI: 10.1039/c0ee00541j. 15

[60] J. Wu, A. E. Contreras, and Q. Li. Studying the impact of RO membrane surface functional groups on alginate fouling in seawater desalination. *J. Memb. Sci.*, 2014. DOI: 10.1016/j.memsci.2014.01.056. 15

[61] E. M. Vrijenhoek, S. Hong, and M. Elimelech. Influence of membrane surface properties on initial rate of colloidal fouling of reverse osmosis and nanofiltration membranes. *J. Memb. Sci.*, 2001. DOI: 10.1016/s0376-7388(01)00376-3. 15, 20

[62] J. S. Louie, I. Pinnau, I. Ciobanu, K. P. Ishida, A. Ng, and M. Reinhard. Effects of polyetherpolyamide block copolymer coating on performance and fouling of reverse osmosis membranes. *J. Memb. Sci.*, 280:762–770, 2006. DOI: 10.1016/j.memsci.2006.02.041. 15

[63] T. Ishigami, K. Amano, A. Fujii, Y. Ohmukai, E. Kamio, T. Maruyama, and H. Matsuyama. Fouling reduction of reverse osmosis membrane by surface modification via layer-by-layer assembly. *Sep. Purif. Technol.*, 99:1–7, 2012. DOI: 10.1016/j.seppur.2012.08.002. 15

[64] A. Rahardianto, W. Y. Shih, R. W. Lee, and Y. Cohen. Diagnostic characterization of gypsum scale formation and control in RO membrane desalination of brackish water. *J. Memb. Sci.*, 2006. DOI: 10.1016/j.memsci.2005.12.059. 15

[65] E. M. V. Hoek, S. Bhattacharjee, and M. Elimelech. Effect of membrane surface roughness on colloid-membrane DLVO interactions. *Langmuir*, 19:4836–4847, 2003. DOI: 10.1021/la027083c. 15, 17

[66] X. Zhu and M. Elimelech. Colloidal fouling of reverse osmosis membranes: Measurements and fouling mechanisms. *Environ. Sci. Technol.*, 1997. DOI: 10.1021/es970400v. 15, 17, 20

[67] G. Z. Ramon and E. M. V. Hoek. Transport through composite membranes, part 2: Impacts of roughness on permeability and fouling. *J. Memb. Sci.*, pages 425–426 and pages 141–148, 2013. DOI: 10.1016/j.memsci.2012.08.004. 15

[68] S. Jiang, Y. Li, and B. P. Ladewig. Science of the total environment: A review of reverse osmosis membrane fouling and control strategies. *Sci. Total Environ.*, 595:567–583, 2017. DOI: 10.1016/j.scitotenv.2017.03.235. 17, 18, 52

[69] S. Jiang, Y. Li, and B. P. Ladewig. A review of reverse osmosis membrane fouling and control strategies. *Sci. Total Environ.*, 595:567–583, 2017. DOI: 10.1016/j.scitotenv.2017.03.235. 15

[70] J. Saqib and I. H. Aljundi. Membrane fouling and modification using surface treatment and layer-by-layer assembly of polyelectrolytes: State-of-the-art review. *J. Water Process Eng.*, 11:68–87, 2016. DOI: 10.1016/j.jwpe.2016.03.009. 18

[71] E. K. Lee, V. Chen, and A. G. Fane. Natural organic matter (NOM) fouling in low pressure membrane filtration—effect of membranes and operation modes. *Desalination*, 218:257–270, 2008. DOI: 10.1016/j.desal.2007.02.021. 15, 18

[72] T. Tong, S. Zhao, C. Boo, S. M. Hashmi, and M. Elimelech. Relating silica scaling in reverse osmosis to membrane surface properties. *Environ. Sci. Technol.*, 51:4396–4406, 2017. DOI: 10.1021/acs.est.6b06411. 15, 59

[73] E. M. V. Hoek. MF/UF membrane filtration: A state-of-the-art review, 2017. www.waterplanet.com 15, 18, 49

[74] L. Malaeb and G. M. Ayoub. Reverse osmosis technology for water treatment: State-of-the-art review. *Desalination*, 267:1–8, 2011. DOI: 10.1016/j.desal.2010.09.001. 15, 18, 49

[75] S. Jiang, Y. Li, and B. P. Ladewig. A review of reverse osmosis membrane fouling and control strategies. *Sci. Total Environ.*, 595:567–583, 2017. DOI: 10.1016/j.scitotenv.2017.03.235. 15, 19

[76] D. Norberg, S. Hong, J. Taylor, and Y. Zhao. Surface characterization and performance evaluation of commercial fouling resistant low-pressure RO membranes. *Desalination*, 202:45–52, 2007. DOI: 10.1016/j.desal.2005.12.037.

[77] G. Kang, H. Yu, Z. Liu, and Y. Cao. Surface modification of a commercial thin film composite polyamide reverse osmosis membrane by carbodiimide-induced grafting with poly(ethylene glycol) derivatives. *Desalination*, 275:252–259, 2011. DOI: 10.1016/j.desal.2011.03.007. 19

[78] K. Y. Jee, D. H. Shin, and Y. T. Lee. Surface modification of polyamide RO membrane for improved fouling resistance. *Desalination*, 394:131–137, 2016. DOI: 10.1016/j.desal.2016.05.013.

[79] Y. jie Xie, H. Yin Yu, S. Yuan Wang, and Z. Kang Xu. Improvement of antifouling characteristics in a bioreactor of polypropylene microporous membrane by the adsorption of Tween 20. *J. Environ. Sci.*, 2007. DOI: 10.1016/s1001-0742(07)60238-1. 18

[80] M. C. Wilbert, J. Pellegrino, and A. Zydney. Bench-scale testing of surfactant-modified reverse osmosis/nanofiltration membranes. *Desalination*, 1998. DOI: 10.1016/s0011-9164(98)00022-8.

[81] Y. Zhou, S. Yu, C. Gao, and X. Feng. Surface modification of thin film composite polyamide membranes by electrostatic self deposition of polycations for improved fouling resistance. *Sep. Purif. Technol.*, 66:287–294, 2009. DOI: 10.1016/j.seppur.2008.12.021.

[82] C. Ba and J. Economy. Preparation and characterization of a neutrally charged antifouling nanofiltration membrane by coating a layer of sulfonated poly(ether ether ketone) on a positively charged nanofiltration membrane. *J. Memb. Sci.*, 2010. DOI: 10.1016/j.memsci.2010.06.032.

[83] H. Hachisuka, P. E. W. L. Walker, A. Examiner, and M. S. Ocampo. Composite reverse osmosis membrane having a separation layer with polyvinyl alcohol coating and method of reverse osmotic treatment of water using the same, 2001. https://patents.google.com/patent/US6177011B1/en. 19

[84] J. S. Louie, I. Pinnau, and M. Reinhard. Effects of surface coating process conditions on the water permeation and salt rejection properties of composite polyamide reverse osmosis membranes. *J. Memb. Sci.*, 2011. DOI: 10.1016/j.memsci.2010.10.067.

[85] A. Sarkar, P. I. Carver, T. Zhang, A. Merrington, K. J. Bruza, J. L. Rousseau, S. E. Keinath, and P. R. Dvornic. Dendrimer-based coatings for surface modification of polyamide reverse osmosis membranes. *J. Memb. Sci.*, 2010. DOI: 10.1016/j.memsci.2009.12.005. 19

[86] A. C. Sagle, E. M. Van Wagner, H. Ju, B. D. McCloskey, B. D. Freeman, and M. M. Sharma. PEGcoated reverse osmosis membranes: Desalination properties and fouling resistance. *J. Memb. Sci.*, 2009. DOI: 10.1016/j.memsci.2009.05.013. 19

[87] E. S. Hatakeyama, H. Ju, C. J. Gabriel, J. L. Lohr, J. E. Bara, R. D. Noble, B. D. Freeman, and D. L. Gin. New protein-resistant coatings for water filtration membranes based on quaternary ammonium and phosphonium polymers. *J. Memb. Sci.*, 2009. DOI: 10.1016/j.memsci.2008.12.049. 19

[88] Y. H. La, B. D. McCloskey, R. Sooriyakumaran, A. Vora, B. Freeman, M. Nassar, J. Hedrick, A. Nelson, and R. Allen. Bifunctional hydrogel coatings for water purification membranes: Improved fouling resistance and antimicrobial activity. *J. Memb. Sci.*, 2011. DOI: 10.1016/j.memsci.2011.02.005. 19

[89] C. Y. Tang, Y. N. Kwon, and J. O. Leckie. Probing the nano- and micro-scales of reverse osmosis membranes—a comprehensive characterization of physiochemical properties of uncoated and coated membranes by XPS, TEM, ATR-FTIR, and streaming potential measurements. *J. Memb. Sci.*, 287:146–156, 2007. DOI: 10.1016/j.memsci.2006.10.038. 19

[90] C. Y. Tang, Y. N. Kwon, and J. O. Leckie. Effect of membrane chemistry and coating layer on physiochemical properties of thin film composite polyamide RO and NF membranes. I. FTIR and XPS characterization of polyamide and coating layer chemistry. *Desalination*, 2009. DOI: 10.1016/j.desal.2008.04.003. 19

[91] C. Y. Tang, Y. N. Kwon, and J. O. Leckie. Effect of membrane chemistry and coating layer on physiochemical properties of thin film composite polyamide RO and NF membranes. II. Membrane physiochemical properties and their dependence on polyamide and coating layers. *Desalination*, 242:168–182, 2009. DOI: 10.1016/j.desal.2008.04.004. 19

[92] L. Zou, I. Vidalis, D. Steele, A. Michelmore, S. P. Low, and J. Q. J. C. Verberk. Surface hydrophilic modification of RO membranes by plasma polymerization for low organic fouling. *J. Memb. Sci.*, 2011. DOI: 10.1016/j.memsci.2010.12.023. 20

[93] X. Wei, Z. Wang, Z. Zhang, J. Wang, and S. Wang. Surface modification of commercial aromatic polyamide reverse osmosis membranes by graft polymerization of 3-allyl-5, 5-dimethylhydantoin. *J. Memb. Sci.*, 351:222–233, 2010. DOI: 10.1016/j.memsci.2010.01.054. 20

[94] A. J. Atkinson, J. Wang, Z. Zhang, A. Gold, D. Jung, D. Zeng, A. Pollard, and O. Coronell. Grafting of bioactive 2-aminoimidazole into active layer makes commercial RO/NF membranes antibiofouling. *J. Memb. Sci.*, 556:85–97, 2018. DOI: 10.1016/j.memsci.2018.03.044. 20

[95] R. Yang, J. Xu, G. Ozaydin-Ince, S. Y. Wong, and K. K. Gleason. Surface-tethered zwitterionic ultrathin antifouling coatings on reverse osmosis membranes by initiated chemical vapor deposition. *Chem. Mater.*, 2011. DOI: 10.1021/cm1031392. 20

[96] C. K. Kim, J. H. Kim, I. J. Roh, and J. J. Kim. The changes of membrane performance with polyamide molecular structure in the reverse osmosis process. *J. Memb. Sci.*, 2000. DOI: 10.1016/s0376-7388(99)00232-x. 21

[97] L. Li, S. Zhang, X. Zhang, and G. Zheng. Polyamide thin film composite membranes prepared from isomeric biphenyl tetraacyl chloride and m-phenylenediamine. *J. Memb. Sci.*, 315:20–27, 2008. DOI: 10.1016/j.memsci.2008.02.022. 21

[98] D. Mukherjee, A. Kulkarni, and W. N. Gill. Chemical treatment for improved performance of reverse osmosis membranes. *Desalination*, 1996. DOI: 10.1016/0011-9164(96)00047-1. 21

[99] H. Karkhanechi, R. Takagi, and H. Matsuyama. Biofouling resistance of reverse osmosis membrane modified with polydopamine. *Desalination*, 336:87–96, 2014. DOI: 10.1016/j.desal.2013.12.033.

[100] C. Zhou, D. Ye, H. Jia, S. Yu, M. Liu, and C. Gao. Surface mineralization of commercial thinfilm composite polyamide membrane by depositing barium sulfate for improved reverse osmosis performance and antifouling property. *Desalination*, 351:228–235, 2014. DOI: 10.1016/j.desal.2014.07.040.

[101] J. Nikkola, X. Liu, Y. Li, M. Raulio, H.-L. Alakomi, J. Wei, and C. Y. Tang. Surface modification of thin film composite RO membrane for enhanced anti-biofouling performance. *J. Memb. Sci.*, 444:192–200, 2013. DOI: 10.1016/j.memsci.2013.05.032.

[102] Y. Zhang, Y. Wan, G. Pan, H. Shi, H. Yan, J. Xu, M. Guo, Z. Wang, and Y. Liu. Surface modification of polyamide reverse osmosis membrane with sulfonated polyvinyl alcohol for antifouling. *Appl. Surf. Sci.*, 419:177–187, 2017. DOI: 10.1016/j.apsusc.2017.05.047.

[103] H. Choi, J. Park, T. Tak, and Y.-N. Kwon. Surface modification of seawater reverse osmosis (SWRO) membrane using methyl methacrylate-hydroxy poly (oxyethylene) methacrylate (MMA-HPOEM) comb-polymer and its performance. *Desalination*, 291:1–7, 2012. DOI: 10.1016/j.desal.2012.01.018.

[104] H. Z. Shafi, A. Matin, Z. Khan, A. Khalil, and K. K. Gleason. Surface modification of reverse osmosis membranes with zwitterionic coatings: A potential strategy for control of biofouling. *Surf. Coatings Technol.*, 279:171–179, 2015. DOI: 10.1016/j.surfcoat.2015.08.037.

[105] R. Yang, H. Jang, R. Stocker, and K. K. Gleason. Synergistic prevention of biofouling in seawater desalination by zwitterionic surfaces and low-level chlorination. *Adv. Mater.*, 26:1711–1718, 2014. DOI: 10.1002/adma.201304386.

[106] S. Azari and L. Zou. Using zwitterionic amino acid l-DOPA to modify the surface of thin film composite polyamide reverse osmosis membranes to increase their fouling resistance. *J. Memb. Sci.*, 401:68–75, 2012. DOI: 10.1016/j.memsci.2012.01.041.

[107] A. Matin, Z. Khan, K. K. Gleason, M. Khaled, S. M. J. Zaidi, A. Khalil, P. Moni, and R. Yang. Surface-modified reverse osmosis membranes applying a copolymer film to reduce adhesion of bacteria as a strategy for biofouling control. *Sep. Purif. Technol.*, 124:117–123, 2014. DOI: 10.1016/j.seppur.2013.12.032.

[108] J. Nikkola, J. Sievänen, M. Raulio, J. Wei, J. Vuorinen, and C. Y. Tang. Surface modification of thin film composite polyamide membrane using atomic layer deposition method. *J. Memb. Sci.*, 450:174–180, 2014. DOI: 10.1016/j.memsci.2013.09.005.

[109] S. Yu, Z. Lü, Z. Chen, X. Liu, M. Liu, and C. Gao. Surface modification of thin-film composite polyamide reverse osmosis membranes by coating N-isopropylacrylamide-co-acrylic acid copolymers for improved membrane properties. *J. Memb. Sci.*, 371:293–306, 2011. DOI: 10.1016/j.memsci.2011.01.059.

[110] Y. Wang, Z. Wang, X. Han, J. Wang, and S. Wang. Improved flux and anti-biofouling performances of reverse osmosis membrane via surface layer-by-layer assembly. *J. Memb. Sci.*, 539:403–411, 2017. DOI: 10.1016/j.memsci.2017.06.029. 21

[111] S. Yu, G. Yao, B. Dong, H. Zhu, X. Peng, J. Liu, M. Liu, and C. Gao. Improving fouling resistance of thin-film composite polyamide reverse osmosis membrane by coating natural hydrophilic polymer sericin. *Sep. Purif. Technol.*, 118:285–293, 2013. DOI: 10.1016/j.seppur.2013.07.018.

[112] F. Shao, L. Dong, H. Dong, Q. Zhang, M. Zhao, L. Yu, B. Pang, and Y. Chen. Graphene oxide modified polyamide reverse osmosis membranes with enhanced chlorine resistance. *J. Memb. Sci.*, 525:9–17, 2017. DOI: 10.1016/j.memsci.2016.12.001.

[113] Y.-N. Kwon, S. Hong, H. Choi, and T. Tak. Surface modification of a polyamide reverse osmosis membrane for chlorine resistance improvement. *J. Memb. Sci.*, 415:192–198, 2012. DOI: 10.1016/j.memsci.2012.04.056. 21

[114] Y. Hu, K. Lu, F. Yan, Y. Shi, P. Yu, S. Yu, S. Li, and C. Gao. Enhancing the performance of aromatic polyamide reverse osmosis membrane by surface modification via covalent attachment of polyvinyl alcohol (PVA). *J. Memb. Sci.*, 501:209–219, 2016. DOI: 10.1016/j.memsci.2015.12.003. 21

[115] T. Zhang, C. Zhu, H. Ma, R. Li, B. Dong, Y. Liu, and S. Li. Surface modification of APA-TFC membrane with quaternary ammonium cation and salicylaldehyde to improve performance. *J. Memb. Sci.*, 457:88–94, 2014. DOI: 10.1016/j.memsci.2014.01.024. 21

[116] M. Kim, N. H. Lin, G. T. Lewis, and Y. Cohen. Surface nano-structuring of reverse osmosis membranes via atmospheric pressure plasma-induced graft polymerization for reduction of mineral scaling propensity. *J. Memb. Sci.*, 354:142–149, 2010. DOI: 10.1016/j.memsci.2010.02.053. 21

[117] A. R. Da Costa and A. G. Fane. Net-type spacers: Effect of configuration on fluid flow path and ultrafiltration flux. *Ind. Eng. Chem. Res.*, 33:1845–1851, 1994. DOI: 10.1021/ie00031a026. 21

[118] A. H. Haidari, S. G. J. Heijman, and W. G. J. van der Meer, Optimal design of spacers in reverse osmosis. *Sep. Purif. Technol.*, 192:441–456, 2018. DOI: 10.1016/j.seppur.2017.10.042. 21, 24, 25

[119] P. R. Neal, H. Li, A. G. Fane, and D. E. Wiley. The effect of filament orientation on critical flux and particle deposition in spacer-filled channels. *J. Memb. Sci.*, 214:165–178, 2003. DOI: 10.1016/s0376-7388(02)00500-8. 21

[120] Y. C. Kim and M. Elimelech. Adverse impact of feed channel spacers on the performance of pressure retarded osmosis. *Environ. Sci. Technol.*, 2012. DOI: 10.1021/es3002597.

[121] T. Tran, B. Bolto, S. Gray, M. Hoang, and E. Ostarcevic. An autopsy study of a fouled reverse osmosis membrane element used in a brackish water treatment plant. *Water Res.*, 2007. DOI: 10.1016/j.watres.2007.06.008.

[122] C. C. Zimmerer and V. Kottke. Effects of spacer geometry on pressure drop, mass transfer, mixing behavior, and residence time distribution. *Desalination*, 1996. DOI: 10.1016/0011-9164(96)00035-5.

[123] J. S. Vrouwenvelder, J. A. M. van Paassen, L. P. Wessels, A. F. van Dam, and S. M. Bakker. The membrane fouling simulator: A practical tool for fouling prediction and control. *J. Memb. Sci.*, 2006. DOI: 10.1016/j.memsci.2006.03.046. 21

[124] S. Sourirajan and J. P. Agrawal. Reverse osmosis, 2005. DOI: 10.1021/ie50719a007. 24

[125] F. Li, W. Meindersma, A. B. De Haan, and T. Reith. Novel spacers for mass transfer enhancement in membrane separations. *J. Memb. Sci.*, 253:1–12, 2005. DOI: 10.1016/j.memsci.2004.12.019. 24, 25

[126] O. Kavianipour, G. D. Ingram, and H. B. Vuthaluru. Investigation into the effectiveness of feed spacer configurations for reverse osmosis membrane modules using computational fluid dynamics. *J. Memb. Sci.*, 2017. DOI: 10.1016/j.memsci.2016.12.034. 25

[127] A. Saeed, R. Vuthaluru, Y. Yang, and H. B. Vuthaluru. Effect of feed spacer arrangement on flow dynamics through spacer filled membranes. *Desalination*, 2012. DOI: 10.1016/j.desal.2011.09.050. 25

[128] F. Li, W. Meindersma, A. B. De Haan, and T. Reith. Optimization of commercial net spacers in spiral wound membrane modules. *J. Memb. Sci.*, 2002. DOI: 10.1016/s0376-7388(02)00307-1. 25

[129] A. Ronen, S. Lerman, G. Z. Ramon, and C. G. Dosoretz. Experimental characterization and numerical simulation of the anti-biofuling activity of nanosilver-modified feed spacers in membrane filtration. *J. Memb. Sci.*, 2015. DOI: 10.1016/j.memsci.2014.10.042. 25

[130] A. Siddiqui, S. Lehmann, S. S. Bucs, M. Fresquet, L. Fel, E. I. E. C. Prest, J. Ogier, C. Schellenberg, M. C. M. van Loosdrecht, J. C. Kruithof, and J. S. Vrouwenvelder. Predicting the impact of feed spacer modification on biofouling by hydraulic characterization and biofouling studies in membrane fouling simulators. *Water Res.*, 2017. DOI: 10.1016/j.watres.2016.12.034. 25

[131] W. S. Tan, S. R. Suwarno, J. An, C. K. Chua, A. G. Fane, and T. H. Chong. Comparison of solid, liquid and powder forms of 3D printing techniques in membrane spacer fabrication. *J. Memb. Sci.*, 2017. DOI: 10.1016/j.memsci.2017.05.037. 25

[132] H. S. Abid, B. S. Lalia, P. Bertoncello, R. Hashaikeh, B. Clifford, D. T. Gethin, and N. Hilal. Electrically conductive spacers for self-cleaning membrane surfaces via periodic electrolysis. *Desalination*, 2017. DOI: 10.1016/j.desal.2017.04.018.

[133] Aqua Membranes, Three-dimensional Printed Spacer Technology, (n.d.). https://aquamembranes.com/technology/

[134] R. Hausman, T. Gullinkala, and I. C. Escobar. Development of low-biofouling polypropylene feedspacers for reverse osmosis. *J. Appl. Polym. Sci.*, 2009. DOI: 10.1002/app.30755.

[135] H. L. Yang, J. C. Te Lin, and C. Huang. Application of nanosilver surface modification to RO membrane and spacer for mitigating biofouling in seawater desalination. *Water Res.*, 2009. DOI: 10.1016/j.watres.2009.06.002.

[136] P. A. Araújo, D. J. Miller, P. B. Correia, M. C. M. Van Loosdrecht, J. C. Kruithof, B. D. Freeman, D. R. Paul, and J. S. Vrouwenvelder. Impact of feed spacer and membrane modification by hydrophilic, bactericidal and biocidal coating on biofouling control. *Desalination*, 295:1–10, 2012. DOI: 10.1016/j.desal.2012.02.026.

[137] P. A. Araújo, J. C. Kruithof, M. C. M. Van Loosdrecht, and J. S. Vrouwenvelder. The potential of standard and modified feed spacers for biofouling control. *J. Memb. Sci.*, pages 403–404 and pages 58–70, 2012. DOI: 10.1016/j.memsci.2012.02.015.

[138] A. Ronen, R. Semiat, and C. G. Dosoretz. Impact of ZnO embedded feed spacer on biofilm development in membrane systems. *Water Res.*, 2013. DOI: 10.1016/j.watres.2013.08.036.

[139] M. Amokrane, D. Sadaoui, M. Dudeck, and C. P. Koutsou. New spacer designs for the performance improvement of the zigzag spacer configuration in spiral-wound membrane modules. *Desalin. Water Treat.*, 2016. DOI: 10.1080/19443994.2015.1022003. 30

[140] J. Balster, I. Pünt, D. F. Stamatialis, and M. Wessling. Multi-layer spacer geometries with improved mass transport. *J. Memb. Sci.*, 2006. DOI: 10.1016/j.memsci.2006.05.039.

[141] A. R. Da Costa, A. G. Fane, C. J. D. Fell, and A. C. M. Franken. Optimal channel spacer design for ultrafiltration. *J. Memb. Sci.*, 1991. DOI: 10.1016/0376-7388(91)80043-6. 38, 69

[142] R. Franks, C. Bartels, and A. Anit. Demonstrating improved RO system performance with new low differential (LD) technology, pages 1–8, 2015.

[143] J. Schwinge, D. E. Wiley, and A. G. Fane. Novel spacer design improves observed flux. *J. Memb. Sci.*, 2004. DOI: 10.1016/j.memsci.2003.09.015.

[144] C. P. Koutsou, S. G. Yiantsios, and A. J. Karabelas. Direct numerical simulation of flow in spacer-filled channels: Effect of spacer geometrical characteristics. *J. Memb. Sci.*, 2007. DOI: 10.1016/j.memsci.2006.12.032.

[145] M. A. Alghoul, P. Poovanaesvaran, K. Sopian, and M. Y. Sulaiman. Review of brackish water reverse osmosis (BWRO) system designs. *Renew. Sustain. Energy Rev.*, 13:2661–2667, 2009. DOI: 10.1016/j.rser.2009.03.013. 29

[146] M. Elimelech and W. A. Phillip. The future of seawater desalination: Energy, technology, and the environment, science, 333:712–718, 2011. DOI: 10.1126/science.1200488.

[147] K. L. Tu, L. D. Nghiem, and A. R. Chivas. Boron removal by reverse osmosis membranes in seawater desalination applications. *Sep. Purif. Technol.*, 75:87–101, 2010. DOI: 10.1016/j.seppur.2010.07.021. 30

[148] K. L. Tu, L. D. Nghiem, and A. R. Chivas. Boron removal by reverse osmosis membranes in seawater desalination applications. *Sep. Purif. Technol.*, 75:87–101, 2010. DOI: 10.1016/j.seppur.2010.07.021. 30

[149] L. A. Richards, M. Vuachère, and A. I. Schäfer. Impact of pH on the removal of fluoride, nitrate and boron by nanofiltration/reverse osmosis. *Desalination*, 261:331–337, 2010. DOI: 10.1016/j.desal.2010.06.025. 30

[150] D. Hasson, H. Shemer, I. Brook, I. Zaslavschi, R. Semiat, C. Bartels, and M. Wilf. Scaling propensity of seawater in RO boron removal processes. *J. Memb. Sci.*, 384:198–204, 2011. DOI: 10.1016/j.memsci.2011.09.027. 30

[151] W. V. Collentro and A. W. Collentro. Purification of gases from water using reverse osmosis, 1997.

[152] X. Jin, A. Jawor, S. Kim, and E. M. V. Hoek. Effects of feed water temperature on separation performance and organic fouling of brackish water RO membranes. *Desalination*, 239:346–359, 2009. DOI: 10.1016/j.desal.2008.03.026. 32

[153] J. M. Edward Sylvester Jr. 5 key performance indicators in reverse osmosis, pages 1–13, 2019.

[154] M. Li. Reducing specific energy consumption in reverse osmosis (RO) water desalination: An analysis from first principles. *Desalination*, 276:128–135, 2011. DOI: 10.1016/j.desal.2011.03.031. 33

[155] L. Song and K. G. Tay. Performance prediction of a long crossflow reverse osmosis membrane channel. *J. Memb. Sci.*, 281:163–169, 2006. DOI: 10.1016/j.memsci.2006.03.026. 33, 76

[156] T. Manth, M. Gabor, and E. Oklejas Jr. Minimizing RO energy consumption under variable conditions of operation. *Desalination*, 157:9–21, 2003. DOI: 10.1016/s0011-9164(03)00377-1. 33

[157] C. J. Gabelich, M. D. Williams, A. Rahardianto, J. C. Franklin, and Y. Cohen. High-recovery reverse osmosis desalination using intermediate chemical demineralization. *J. Memb. Sci.*, 301:131–141, 2007. DOI: 10.1016/j.memsci.2007.06.007. 33

[158] E. Tzen and R. Morris. Renewable energy sources for desalination. *Sol. Energy*, 75:375–379, 2003. DOI: 10.1016/j.solener.2003.07.010. 33, 34

[159] H.-J. Oh, T.-M. Hwang, and S. Lee. A simplified simulation model of RO systems for seawater desalination. *Desalination*, 238:128–139, 2009. DOI: 10.1016/j.desal.2008.01.043. 33

[160] P. Geisler, F. U. Hahnenstein, W. Krumm, and T. Peters. Pressure exchange system for energy recovery in reverse osmosis plants. *Desalination*, 122:151–156, 1999. DOI: 10.1016/s0011-9164(99)00036-3. 34

[161] S. A. Avlonitis, K. Kouroumbas, and N. Vlachakis. Energy consumption and membrane replacement cost for seawater RO desalination plants. *Desalination*, 157:151–158, 2003. DOI: 10.1016/s0011-9164(03)00395-3. 34

[162] M. Li. Reducing specific energy consumption in Reverse Osmosis (RO) water desalination: An analysis from first principles. *Desalination*, 276:128–135, 2011. DOI: 10.1016/j.desal.2011.03.031. 34

[163] F. L. Gordon. Breaking the cost barrier for seawater desalting. *Int. Desalin. and Water Reuse.*, 8:14–20, 1998. 34

[164] S. A. Avlonitis, K. Kouroumbas, and N. Vlachakis. Energy consumption and membrane replacement cost for seawater RO desalination plants. *Desalination*, 157:151–158, 2003. DOI: 10.1016/s0011-9164(03)00395-3. 34, 39

[165] M. Thomson, M. S. Miranda, and D. Infield. A small-scale seawater reverse-osmosis system with excellent energy efficiency over a wide operating range. *Desalination*, 153:229–236, 2003. DOI: 10.1016/s0011-9164(02)01141-4. 34, 37

[166] B. Peñate and L. García-Rodríguez. Current trends and future prospects in the design of seawater reverse osmosis desalination technology. *Desalination*, 284:1–8, 2012. DOI: 10.1016/j.desal.2011.09.010. 34

[167] F. Banat. Economic and technical assessment of desalination technologies. *IWA Conf. Technol. Water ...*, 2007. http://www.desline.com/Geneva/Banat.pdf 34, 37

[168] N. Voutchkov. Desalination project cost estimating and management, 2019. https://doi.org/10.1201/9781351242738. DOI: 10.1201/9781351242738. 35, 36, 37, 38

[169] R. S. El-Emam and I. Dincer. Thermodynamic and thermoeconomic analyses of seawater reverse osmosis desalination plant with energy recovery. *Energy*, 64:154–163, 2014. DOI: 10.1016/j.energy.2013.11.037. 38

[170] G. E. Gruen. Turkish waters: Source of regional conflict or catalyst for peace? *Environ. Challenges*, pages 565–579, Springer, 2000. DOI: 10.1007/978-94-011-4369-1_44.

[171] I. C. Karagiannis and P. G. Soldatos. Water desalination cost literature: Review and assessment. *Desalination*, 223:448–456, 2008. DOI: 10.1016/j.desal.2007.02.071. 38

[172] S. H. Ebrahim, M. M. Abdel-Jawad, and M. Safar. Conventional pre-treatment system for the doha reverse osmosis plant: Technical and economic assessment. *Desalination*, 102:179–187, 1995. DOI: 10.1016/0011-9164(95)00052-4. 54

[173] F. Knops, S. van Hoof, H. Futselaar, and L. Broens. Economic evaluation of a new ultra-filtration membrane for pre-treatment of seawater reverse osmosis. *Desalination*, 203:300–306, 2007. DOI: 10.1016/j.desal.2006.04.013. 38

[174] P. Glueckstern and M. Priel. Comparative cost of UF vs. conventional pre-treatment for SWRO systems. *Int. Desalin. Water Reuse Q.*, 13:34–39, 2003.

[175] A. Pérez-González, A. M. Urtiaga, R. Ibáñez, and I. Ortiz. State-of-the-art and review on the treatment technologies of water reverse osmosis concentrates. *Water Res.*, 46:267–283, 2012. DOI: 10.1016/j.watres.2011.10.046.

[176] B. D. Stanford, J. F. Leising, R. G. Bond, and S. A. Snyder. Inland desalination: Current practices, environmental implications, and case studies in Las Vegas, NV. *Sustain. Sci. Eng.*, 2:327–350, 2010. DOI: 10.1016/s1871-2711(09)00211-6. 38

[177] M. Ahmed, W. H. Shayya, D. Hoey, and J. Al-Handaly. Brine disposal from reverse osmosis desalination plants in Oman and the United Arab Emirates. *Desalination*, 133:135–147, 2001. DOI: 10.1016/s0011-9164(01)80004-7. 38

[178] J. M. Arnal, M. Sancho, I. Iborra, J. M. Gozálvez, A. Santafé, and J. Lora. Concentration of brines from RO desalination plants by natural evaporation. *Desalination*, 182:435–439, 2005. DOI: 10.1016/j.desal.2005.02.036. 38

[179] Y. Lu, Y. Hu, D. Xu, and L. Wu. Optimum design of reverse osmosis seawater desalination system considering membrane cleaning and replacing. *J. Memb. Sci.*, 282:7–13, 2006. DOI: 10.1016/j.memsci.2006.04.019. 38

[180] M. Ahmed, W. H. Shayya, D. Hoey, A. Mahendran, R. Morris, and J. Al-Handaly. Use of evaporation ponds for brine disposal in desalination plants. *Desalination*, 130:155–168, 2000. DOI: 10.1016/s0011-9164(00)00083-7.

[181] S. Masnoon and K. Glucina. Desalination: Brine and residual management. *Water Res. Found.*, 2011. 39, 41

[182] P. Malmrose, J. Lozier, M. Mickley, and R. Reiss. Committee report: Current perspectives on residuals management for desalting membranes. *Am. Water Work. Assoc. J.*, 96:73, 2004. DOI: 10.1002/j.1551-8833.2004.tb10760.x. 42, 116

[183] B. Alspach. Conventional ZLD: Overview and economics. *AMTA/SWMOA Jt. Technol. Transf. Work. Enhanced Recover. NF/RO Syst.*, Albuquerque, NM, 2019. 43, 111, 112, 113

[184] G. Juby. Evaluation and selection of available processes for a zero-liquid discharge system for the Perris, California, ground water basin, 2008. https://www.usbr.gov/research/dwpr/reportpdfs/report149.pdf 61

[185] R. Stover. Counter Flow RO (CFRO) membrane brine concentration. *AMTASWMOA Technol. Transf. Work. Enhanced Recover. NF/RO Syst.*, Albuquerque, NM, 2019. 44, 45, 94, 95

[186] B. Alspach and G. Juby. Cost-effective ZLD technology for desalination concentrate management. *J. Am. Water Work. Assoc.*, 110, 2018. DOI: 10.5942/jawwa.2018.110.0005. 45, 46

[187] P. H. Wolf, S. Siverns, and S. Monti. UF membranes for RO desalination pre-treatment. *Desalination*, 182:293–300, 2005. DOI: 10.1016/j.desal.2005.05.006. 49

[188] H. Huang and K. Schwab, J. G. Jacangelo. Pretreatment for low pressure membranes in water treatment: A review. *Environ. Sci. Technol.*, 43:3011–3019, 2009. DOI: 10.1021/es802473r. 49, 52, 53

[189] T. Qiu and P. A. Davies. Comparison of configurations for high-recovery inland desalination systems. *Water*, 4:690–706, Switzerland, 2012. DOI: 10.3390/w4030690. 49, 52

[190] S. Jamaly, N. N. Darwish, I. Ahmed, and S. W. Hasan. A short review on reverse osmosis pre-treatment technologies. *Desalination*, 354:30–38, 2014. DOI: 10.1016/j.desal.2014.09.017. 49

[191] A. W. Mohamma, N. A. Darwish, M. Abu Arabi, N. Hilal, and H. Al-Zoubi. A comprehensive review of nanofiltration membranes: Treatment, pre-treatment, modelling, and atomic force microscopy. *Desalination*, 170:281–308, 2005. DOI: 10.1016/j.desal.2004.01.007. 49, 68

[192] S. S. Shenvi, A. M. Isloor, and A. F. Ismail. A review on RO membrane technology: Developments and challenges. *Desalination*, 368:10–26, 2015. DOI: 10.1016/j.desal.2014.12.042. 49

[193] Y.-M. Yoon, J.-S. Sohn, R. Valavala, N.-G. Her, and J.-H. Han. Pretreatment in reverse osmosis seawater desalination: A short review. *Environ. Eng. Res.*, 16:205–212, 2012. DOI: 10.4491/eer.2011.16.4.205. 49, 52

[194] R. Verbeke, V. Gómez, and I. F. J. Vankelecom. Chlorine-resistance of reverse osmosis (RO) polyamide membranes. *Prog. Polym. Sci.*, 72:1–15, 2017. DOI: 10.1016/j.progpolymsci.2017.05.003. 54, 55, 56, 71

[195] W. R. Adams. The effects of chlorine dioxide on reverse osmosis membranes. *Desalination*, 78:439–453, 1990. DOI: 10.1016/0011-9164(90)80063-h. 55

[196] I. C. Tessaro, J. B. A. da Silva, and K. Wada. Investigation of some aspects related to the degradation of polyamide membranes: Aqueous chlorine oxidation catalyzed by aluminum and sodium laurel sulfate oxidation during cleaning. *Desalination*, 181:275–282, 2005. DOI: 10.1016/j.desal.2005.04.008. 55, 71

[197] C. J. Gabelich, J. C. Frankin, F. W. Gerringer, K. P. Ishida, and I. H. Suffet. Enhanced oxidation of polyamide membranes using monochloramine and ferrous iron. *J. Memb. Sci.*, 258:64–70, 2005. DOI: 10.1016/j.memsci.2005.02.034. 55, 71

[198] V. T. Do, C. Y. Tang, M. Reinhard, and J. O. Leckie. Effects of chlorine exposure conditions on physiochemical properties and performance of a polyamide membrane-mechanisms and implications. *Environ. Sci. Technol.*, 46:13184–13192, 2012. DOI: 10.1021/es302867f. 55

[199] M. J. Cran, S. W. Bigger, and S. R. Gray. Degradation of polyamide reverse osmosis membranes in the presence of chloramine. *Desalination*, 283:58–63, 2011. DOI: 10.1016/j.desal.2011.04.050. 55, 71

[200] D. Kim, G. L. Amy, and T. Karanfil. Disinfection by-product formation during seawater desalination: A review. *Water Res.*, 81:343–355, 2015. DOI: 10.1016/j.watres.2015.05.040. 56

[201] Y. Jin, H. Lee, M. Zhan, and S. Hong. UV radiation pre-treatment for reverse osmosis (RO) process in ultrapure water (UPW) production. *Desalination*, 439:138–146, 2018. DOI: 10.1016/j.desal.2018.04.019. 56

[202] S. Viitasalo, J. Sassi, J. Rytkönen, and E. Leppäkoski. Ozone, ultraviolet light, ultrasound and hydrogen peroxide as ballast water treatments—experiments with mesozooplankton in lowsaline brackish water. *J. Mar. Environ. Eng.*, 8:35–55, 2005. https://www.scopus.com/inward/record.uri?eid=2-s2.0-23944514607&partnerID=40&md5=daedd777f64ca8d9a5169716843f1ff4. 56

[203] C. J. Gabelich, T. I. Yun, B. M. Coffey, and I. H. M. Suffet. Effects of aluminum sulfate and ferric chloride coagulant residuals on polyamide membrane performance. *Desalination*, 150:15–30, 2002. DOI: 10.1016/s0011-9164(02)00926-8. 51, 52, 71

[204] Handbook of Industrial Water Treatment, (n.d.). https://www.suezwatertechnologies.com/handbook/chapter-07-precipitation-softening 52, 64, 66

[205] J. C. Crittenden. Water treatment principles and design, 2005. 52, 53

[206] J. J. Qin, H. O. Maung, H. Lee, and R. Kolkman. Dead-end ultrafiltration for pre-treatment of RO in reclamation of municipal wastewater effluent. *J. Memb. Sci.*, 243:107–113, 2004. DOI: 10.1016/j.memsci.2004.06.010. 53

[207] P. Glueckstern, M. Priel, and M. Wilf. Field evaluation of capillary UF technology as a pre-treatment for large seawater RO systems. *Desalination*, 147:55–62, 2002. DOI: 10.1016/s0011-9164(02)00576-3. 53

[208] A. Altaee and N. Hilal. High recovery rate NF-FO-RO hybrid system for inland brackish water treatment. *Desalination*, 363:19–25, 2015. DOI: 10.1016/j.desal.2014.12.017. 53, 69

[209] C. K. Teng, M. N. A. Hawlader, and A. Malek. An experiment with different pre-treatment methods. *Desalination*, 156:51–58, 2003. DOI: 10.1016/s0011-9164(03)00324-2. 53

[210] C. Tzotzi, T. Pahiadaki, S. G. Yiantsios, A. J. Karabelas, and N. Andritsos. A study of $CaCO_3$ scale formation and inhibition in RO and NF membrane processes, 296:171–184, 2007. DOI: 10.1016/j.memsci.2007.03.031. 57, 58

[211] D. Lisitsin, Q. Yang, D. Hasson, and R. Semiat. Inhibition of $CaCO_3$ scaling on RO membranes by trace amounts of zinc ions, 183:289–300, 2009. DOI: 10.1016/j.desal.2005.10.002. 58

[212] X. Li, B. Gao, Q. Yue, D. Ma, H. Rong, P. Zhao, and P. Teng. Effect of six kinds of scale inhibitors on calcium carbonate precipitation in high salinity wastewater at high temperatures. *J. Environ. Sci.*, 29:124–130, China, 2015. DOI: 10.1016/j.jes.2014.09.027. 58

[213] M. Fathy, T. A. Moghny, M. A. Mousa, and A. E. Awadallah. Sulfonated ion exchange polystyrene composite resin for calcium hardness removal, 5:20–29, 2015. 67

[214] J. Wang, H. Li, C. Shuang, A. Li, C. Wang, and Y. Huang. Effect of pore structure on adsorption behavior of ibuprofen by magnetic anion exchange resins. *Microporous Mesoporous Mater.*, 210:94–100, 2015. DOI: 10.1016/j.micromeso.2015.02.026. 67

[215] B. Karabacakoğlu, F. Tezakıl, and A. Güvenç. Removal of hardness by electrodialysis using homogeneous and heterogeneous ion exchange membranes. *Desalin. Water Treat.*, 54:8–14, 2015. DOI: 10.1080/19443994.2014.880159. 69

[216] X. Wang, Y. Nie, X. Zhang, S. Zhang, and J. Li. Recovery of ionic liquids from dilute aqueous solutions by electrodialysis. *Desalination*, 285:205–212, 2012. DOI: 10.1016/j.desal.2011.10.003.

[217] S. J. Seo, H. Jeon, J. K. Lee, G. Y. Kim, D. Park, H. Nojima, J. Lee, and S. H. Moon, Investigation on removal of hardness ions by capacitive deionization (CDI) for water softening applications. *Water Res.*, 44:2267–2275, 2010. DOI: 10.1016/j.watres.2009.10.020. 70

[218] G. Zalmonb. Compat accelerated precipitation softening (CAPS) as a pre-treatment for membrane desalination, 113:65–71, 1997. DOI: 10.1016/S0011-9164(97)00115-X. 65

[219] Y. Oren, V. Katz, and N. C. Daltrophe. Improved compact accelerated precipitation softening (CAPS). *Desalination*, 139:155–159, 2001. DOI: 10.1016/s0011-9164(01)00305-8. 65

[220] J. Gilron, N. Daltrophe, M. Waissman, and Y. Oren. Comparison between compact accelerated precipitation softening (CAPS) and conventional pre-treatment in operation of brackish water reverse osmosis (BWRO). *Ind. Eng. Chem. Res.*, 44:5465–5471, 2005. DOI: 10.1021/ie050002y. 65

[221] J. Gilron, D. Chaikin, and N. Daltrophe. Demonstration of CAPS pre-treatment of surface water for RO. *Desalination*, 127:271–282, 2000. DOI: 10.1016/s0011-9164(00)00016-3. 65

[222] Y. Oren, V. Katz, and N. C. Daltrophe. Compact accelerated precipitation softening (CAPS) with submerged filtration: Role of the $CaCO_3$ cake and the slurry. *Ind. Eng. Chem. Res.*, 41:5308–5315, 2002. DOI: 10.1021/ie010740w. 65

[223] A. Masarwa, D. Meyerstein, N. Daltrophe, and O. Kedem. Compact accelerated precipitation softening (CAPS) as pre-treatment for membrane desalination II. Lime softening with concomitant removal of silica and heavy metals. *Desalination*, 113:73–84, 1997. DOI: 10.1016/s0011-9164(97)00116-1. 65

[224] Boiler Feedwater Treatment (Part II): Water Treatment Fundamentals, (n.d.). http://www.sedifilt.com/technical_library/boiler_feedwater_treatment_part_ii_water_treatment_fundamentals.html 51, 66

[225] S. El-Manharawy and A. Hafez. Water type and guidelines for RO system design. *Desalination*, 139:97–113, 2001. DOI: 10.1016/s0011-9164(01)00298-3. 49

[226] M. A. Alghoul, P. Poovanaesvaran, K. Sopian, and M. Y. Sulaiman. Review of brackish water reverse osmosis (BWRO) system designs. *Renew. Sustain. Energy Rev.*, 13:2661–2667, 2009. DOI: 10.1016/j.rser.2009.03.013. 49, 51

[227] M. D. Afonso, J. O. Jaber, and M. S. Mohsen. Brackish groundwater treatment by reverse osmosis in Jordan. *Desalination*, 164:157–171, 2004. DOI: 10.1016/s0011-9164(04)00175-4. 49

[228] J. Van Agtmaal, P. A. de Boks, R. Cornips, and L. L. M. J. Paping. Evaluation of feed water sources and retrofitting of an integrated membrane system, pages 1–7, 2007. 49

[229] M. Abdel-Jawad, S. Ebrahim, F. Al-Atram, and S. Al-Shammari. Pretreatment of the municipal wastewater feed for reverse osmosis plants. *Desalination*, 109:211–223, 1997. DOI: 10.1016/s0011-9164(97)00066-0. 49

[230] K. Sadeddin, A. Naser, and A. Firas. Removal of turbidity and suspended solids by electrocoagulation to improve feed water quality of reverse osmosis plant. *Desalination*, 268:204–207, 2011. DOI: 10.1016/j.desal.2010.10.027. 49, 51

[231] S. Jamaly, N. N. Darwish, I. Ahmed, and S. W. Hasan. A short review on reverse osmosis pre-treatment technologies, *Desalination*, 354:30–38, 2014. DOI: 10.1016/j.desal.2014.09.017. 51

[232] C. F. Wan, S. Jin, and T. S. Chung. Mitigation of inorganic fouling on pressure retarded osmosis (PRO) membranes by coagulation pre-treatment of the wastewater concentrate feed. *J. Memb. Sci.*, 572:658–667, 2019. DOI: 10.1016/j.memsci.2018.11.051. 51

[233] A. Zouboulis, G. Traskas, and P. Samaras. Comparison of single and dual media filtration in a full-scale drinking water treatment plant. *Desalination*, 213:334–342, 2007. DOI: 10.1016/j.desal.2006.02.102. 52

[234] S. F. Anis, R. Hashaikeh, and N. Hilal. Reverse osmosis pre-treatment technologies and future trends: A comprehensive review. *Desalination*, 452:159–195, 2019. DOI: 10.1016/j.desal.2018.11.006. 52, 53

[235] Toprak, Wastewater Engineering, 2006. http://web.deu.edu.tr/atiksu/ana52/ani4044-13.html 52

[236] A. Múñoz Elguera and S. O. Pérez Báez. Development of the most adequate pre-treatment for high capacity seawater desalination plants with open intake. *Desalination*, 184:173–183, 2005. DOI: 10.1016/j.desal.2005.04.033. 52

[237] L. Henthorne and B. Boysen. State-of-the-art of reverse osmosis desalination pre-treatment. *Desalination*, 356:129–139, 2015. DOI: 10.1016/j.desal.2014.10.039. 52

[238] A. Brehant, V. Bonnelye, and M. Perez. Comparison of MF/UF pre-treatment with conventional filtration prior to RO membranes for surface seawater desalination. *Desalination*, 144:353–360, 2002. DOI: 10.1016/s0011-9164(02)00343-0. 53

[239] A. Brehant, V. Bonnelye, and M. Perez. Assessment of ultrafiltration as a pre-treatment of reverse osmosis membranes for surface seawater desalination. *Water Supply*, 3:437–445, 2003. DOI: 10.2166/ws.2003.0200. 53

[240] G. K. Pearce. The case for UF/MF pre-treatment to RO in seawater applications, 203:286–295, 2007. DOI: 10.1016/j.desal.2006.04.011. 53, 54

[241] C. V. Vedavyasan. Pre-treatment trends—an overview. *Desalination*, 203:296–299, 2007. DOI: 10.1016/j.desal.2006.04.012. 53

[242] V. Bonnélye, L. Guey, and J. Del Castillo. UF/MF as RO pre-treatment: The real benefit, 222:59–65, 2008. DOI: 10.1016/j.desal.2007.01.129. 53

[243] V. Frenkel. Membrane Technologies: State-of-the-art and areas of future innovation. *UCLA Civ. Environ. Eng. Semin.*, 2019. 53

[244] S. Arrojo, Y. Benito, and A. Martínez Tarifa. A parametrical study of disinfection with hydrodynamic cavitation. *Ultrason. Sonochem.*, 15:903–908, 2008. DOI: 10.1016/j.ultsonch.2007.11.001. 54

[245] T. Lichi, E. Margalit, M. Herzberg, T. Harif, D. Minz, and H. Elifantz. The effect of UV pre-treatment on biofouling of BWRO membranes: A field study. *Desalin. Water Treat.*, 31:151–163, 2011. DOI: 10.5004/dwt.2011.2377. 54

[246] C. M. Davies, D. J. Roser, A. J. Feitz, and N. J. Ashbolt. Solar radiation disinfection of drinking water at temperate latitudes: Inactivation rates for an optimised reactor configuration. *Water Res.*, 43:643–652, 2009. DOI: 10.1016/j.watres.2008.11.016. 54

[247] A. Hulsmans, K. Joris, N. Lambert, H. Rediers, P. Declerck, Y. Delaedt, F. Ollevier, and S. Liers. Evaluation of process parameters of ultrasonic treatment of bacterial suspensions in a pilot scale water disinfection system. *Ultrason. Sonochem.*, 17:1004–1009, 2010. DOI: 10.1016/j.ultsonch.2009.10.013. 54

[248] M. I. Kerwick, S. M. Reddy, A. H. L. Chamberlain, and D. M. Holt. Electrochemical disinfection, an environmentally acceptable method of drinking water disinfection? *Electrochim. Acta.*, 50:5270–5277, 2005. DOI: 10.1016/j.electacta.2005.02.074. 54

[249] E. Agus, N. Voutchkov, and D. L. Sedlak. Disinfection by-products and their potential impact on the quality of water produced by desalination systems: A literature review. *Desalination*, 237:214–237, 2009. DOI: 10.1016/j.desal.2007.11.059. 54, 55, 56, 57

[250] K. Yadav, K. Morison, and M. P. Staiger. Effects of hypochlorite treatment on the surface morphology and mechanical properties of polyethersulfone ultrafiltration membranes. *Polym. Degrad. Stab.*, 94:1955–1961, 2009. DOI: 10.1016/j.polymdegradstab.2009.07.027. 54

[251] E. Friedler, I. Katz, and C. G. Dosoretz. Chlorination and coagulation as pre-treatments for greywater desalination. *Desalination*, 222:38–49, 2008. DOI: 10.1016/j.desal.2007.01.130. 54

[252] J. Powell, J. Luh, and O. Coronell. Bulk chlorine uptake by polyamide active layers of thin-film composite membranes upon exposure to free chlorine kinetics, mechanisms, and modeling. *Environ. Sci. Technol.*, 48:2741–2749, 2014. DOI: 10.1021/es4047632. 55

[253] G. Barassi and T. Borrmann. N-chlorination and Orton rearrangement of aromatic polyamides, revisited. *J. Membr. Sci. Technol.*, 2:1000115, 2012. DOI: 10.4172/2155-9589.1000115. 55

[254] H. Li, P. Yu, H. Li, and Y. Luo. The chlorination and chlorine resistance modification of composite polyamide membrane. *J. Appl. Polym. Sci.*, 132, 2015. DOI: 10.1002/app.41584. 55

[255] A. Antony, R. Fudianto, S. Cox, and G. Leslie. Assessing the oxidative degradation of polyamide reverse osmosis membrane-accelerated ageing with hypochlorite exposure. *J. Memb. Sci.*, 347:159–164, 2010. DOI: 10.1016/j.memsci.2009.10.018. 55

[256] J. Powell, J. Luh, and O. Coronell. Amide link scission in the polyamide active layers of thin-film composite membranes upon exposure to free chlorine: Kinetics and mechanisms. *Environ. Sci. Technol.*, 49:12136–12144, 2015. DOI: 10.1021/acs.est.5b02110. 55

[257] J. Xu, Z. Wang, X. Wei, S. Yang, J. Wang, and S. Wang. The chlorination process of crosslinked aromatic polyamide reverse osmosis membrane: New insights from the study of self-made membrane. *Desalination*, 313:145–155, 2013. DOI: 10.1016/j.desal.2012.12.020. 55

[258] R. Singh. Polyamide polymer solution behaviour under chlorination conditions. *J. Memb. Sci.*, 88:285–287, 1994. DOI: 10.1016/0376-7388(94)87015-2. 56

[259] R. E. Larson, J. E. Cadotte, and R. J. Petersen. The FT-30 seawater reverse osmosis membrane—element test results. *Desalination*, 38:473–483, 1981. DOI: 10.1016/s0011-9164(00)86092-0. 56

[260] C. O. Lee, K. J. Howe, and B. M. Thomson. Ozone and biofiltration as an alternative to reverse osmosis for removing PPCPs and micropollutants from treated wastewater. *Water Res.*, 46:1005–1014, 2012. DOI: 10.1016/j.watres.2011.11.069. 56

[261] K. Verma, D. Gupta, and A. B. Gupta. Optimization of ozone disinfection and its effect on trihalomethanes. *J. Environ. Chem. Eng.*, 4:3021–3032, 2016. DOI: 10.1016/j.jece.2016.06.017. 56

[262] M. N. Rojas-Valencia. Research on ozone application as disinfectant and action mechanisms on wastewater microorganisms. *Virus*, 3:0–4, 2011. 56

[263] C. Sigmon, G.-A. Shin, J. Mieog, and K. G. Linden. Establishing surrogate—virus relationships for ozone disinfection of wastewater. *Environ. Eng. Sci.*, 32:451–460, 2015. DOI: 10.1089/ees.2014.0496. 56

[264] P. Paraskeva and N. J. D. Graham. Treatment of a secondary municipal effluent by ozone, UV and microfiltration: Microbial reduction and effect on effluent quality. *Desalination*, 186:47–56, 2005. DOI: 10.1016/j.desal.2005.04.057. 56

[265] G. Carvajal, A. Branch, P. Michel, S. A. Sisson, D. J. Roser, J. E. Drewes, and S. J. Khan. Robust evaluation of performance monitoring options for ozone disinfection in water recycling using Bayesian analysis. *Water Res.*, 124:605–617, 2017. DOI: 10.1016/j.watres.2017.07.079. 56

[266] J. Hotgne and H. Bader. Rate constants of reactions of ozone with organic and inorganic compounds in water-I non-dissociating organic compounds. *Water Re.*, l, 1981. DOI: 10.1016/0043-1354(83)90098-2. 56

[267] M. Vaca, L. Torres, N. Rojas-Valencia, E. Bandala, R. López, and Y. Bustos. Disinfection of primary municipal wastewater effluents using continuous UV and ozone treatment. *J. Water Resour. Prot.*, 06:16–21, 2014. DOI: 10.4236/jwarp.2014.61003. 56

[268] U. Epa. Wastewater technology fact sheet ozone disinfection, 1999. 56

[269] J. Oh, D. E. Salcedo, C. A. Medriano, and S. Kim. Comparison of different disinfection processes in the effective removal of antibiotic-resistant bacteria and genes. *J. Environ. Sci.*, 26:1238–1242, China, 2014. DOI: 10.1016/s1001-0742(13)60594-x. 56

[270] C. Gottschalk, J. A. Libra, and A. Saupe. *Ozonation of Water and Wastewater: A Practical Guide to Understanding Ozone and its Applications*, John Wiley & Sons, 2009. DOI: 10.1002/9783527628926. 56

[271] D. B. Miklos, C. Remy, M. Jekel, K. G. Linden, J. E. Drewes, and U. Hübner. Evaluation of advanced oxidation processes for water and wastewater treatment—A critical review. *Water Res.*, 139:118–131, 2018. DOI: 10.1016/j.watres.2018.03.042. 56

[272] J. Fang, Y. Fu, and C. Shang. The roles of reactive species in micropollutant degradation in the UV/free chlorine system. *Environ. Sci. Technol.*, 48:1859–1868, 2014. DOI: 10.1021/es4036094. 56

[273] M. Processes, R. Rautenbach, and R. Albrezht. *Membrane Processes*, John Wiley & Sons, UK, pages 236–237, 1989, reprinted 1994. 57

[274] Membrane Scale Control, (n.d.). https://www.ecolab.com/nalcowater/offerings/membrane-scale-control 57

[275] R. Ketrane, B. Saidani, O. Gil, L. Leleyter, and F. Baraud. Efficiency of five scale inhibitors on calcium carbonate precipitation from hard water: Effect of temperature and concentration. *Desalination*, 249:1397–1404, 2009. DOI: 10.1016/j.desal.2009.06.013. 58

[276] S. Liakaki-Stavropoulou, M. Danilycheva, A. Andrianov, R. Efremov, and K. D. Demadis. Comparative performance of tetraphosphonate and diphosphonate as reverse osmosis scale inhibitors. *MATEC Web Conf.*, 251:03049, 2018. DOI: 10.1051/matec-conf/201825103049. 58

[277] C. Corporation. Computer modeling of the specific matching between scale inhibitors, 76:57–62, 1986. DOI: 10.1016/0022-0248(86)90008-4. 58

[278] S. A. Ali, I. W. Kazi, and F. Rahman. Synthesis and evaluation of phosphate-free anti-scalants to control $CaSO_4 \cdot 2H_2O$ scale formation in reverse osmosis desalination plants. *Desalination*, 357:36–44, 2015. DOI: 10.1016/j.desal.2014.11.006. 58, 59

[279] Z. Amjad. Applications of antiscalants to control calcium sulfate scaling in reverse osmosis systems. *Desalination*, 54:263–276, 1985. DOI: 10.1016/0011-9164(85)80022-9. 58

[280] A. A. Koelmans, A. Van der Heijde, L. M. Knijff, and R. H. Aalderink. Integrated modelling of eutrophication and organic contaminant fate and effects in aquatic ecosystems. A review. *Water Res.*, 35:3517–3536, 2001. DOI: 10.1016/S0043-1354(01)00095-1. 58

[281] D. Hasson, H. Shemer, and A. Sher. State-of-the-art of friendly green scale control inhibitors: A review article. *Ind. Eng. Chem. Res.*, 50:7601–7607, 2011. DOI: 10.1021/ie200370v. 59

[282] X. Xue, C. Fu, N. Li, F. Zheng, W. Yang, and X. Yang. Performance of a non-phosphorus antiscalant on inhibition of calcium-sulfate precipitation. *Water Sci. Technol.*, 66:193–200, 2012. DOI: 10.2166/wst.2012.157. 59

[283] M. K. Shahid and Y. G. Choi. The comparative study for scale inhibition on surface of RO membranes in wastewater reclamation: CO_2 purging vs. three different antiscalants. *J. Memb. Sci.*, 546:61–69, 2018. DOI: 10.1016/j.memsci.2017.09.087. 59

[284] K. D. Demadis and A. Stathoulopoulou. Solubility enhancement of silicate with polyamine/polyammonium cationic macromolecules: Relevance to silica-laden process waters, pages 4436–4440, 2006. DOI: 10.1021/ie0602254. 59

[285] K. D. Demadis, E. Neofotistou, E. Mavredaki, M. Tsiknakis, E. Sarigiannidou, and S. D. Katarachia. Inorganic foulants in membrane systems: Chemical control strategies and the contribution of "green chemistry," 179:281–295, 2005. DOI: 10.1016/j.desal.2004.11.074. 59

[286] E. Mavredaki, A. Stathoulopoulou, E. Neofotistou, and K. D. Demadis. Environmentally benign chemical additives in the treatment and chemical cleaning of process water systems: Implications for green chemical technology, 210:257–265, 2007. DOI: 10.1016/j.desal.2006.05.050. 59

[287] A. Ketsetzi, A. Stathoulopoulou, and K. D. Demadis. Being green in chemical water treatment technologies: Issues, challenges, and developments, 223:487–493, 2008. DOI: 10.1016/j.desal.2007.01.230. 59

[288] Z. Amjad. Silica scale control by non-ionic polymers: The influence of water system impurities. *Int. J. Corros. Scale Inhib.*, 5:100–111, 2016. DOI: 10.17675/2305-6894-2016-5-2-1. 59, 60

[289] E. Neofotistou and K. D. Demadis. Use of antiscalants for mitigation of silica. *Desalination*, 167:257–272, 2004. DOI: 10.1016/j.desal.2004.06.135. 59, 60

[290] S. D. N. Freeman and R. J. Majerle. Silica fouling revisited. *Desalination*, 103:113–115, 1995. DOI: 10.1016/0011-9164(95)00092-5. 59

[291] S. H. Chan, Z. J. Chen, and P. He. Effect of ferric chloride on silica fouling. *J. Heat Transfer.*, 117:323, 2008. DOI: 10.1115/1.2822524. 59

[292] M. Xie and S. R. Gray. Silica scaling in forward osmosis: From solution to membrane interface. *Water Res.*, 108:232–239, 2017. DOI: 10.1016/j.watres.2016.10.082. 59

[293] B. Mi and M. Elimelech. Silica scaling and scaling reversibility in forward osmosis. *Desalination*, 312:75–81, 2013. DOI: 10.1016/j.desal.2012.08.034. 59

[294] E. Neofotistou and K. D. Demadis. Silica scale inhibition by polyaminoamide STARBURST®dendrimers. *Colloids Surfaces A: Physicochem. Eng. Asp.*, 242:213–216, 2004. DOI: 10.1016/j.colsurfa.2004.04.067. 59, 60

[295] N. Bai, C. Chen, G. Chen, Y. Zhang, M. Xia, and L. Jiao. A theoretical study of the inhibition effect of PAMAM molecule on silica scale. *J. Mol. Model.*, 23:0–5, 2017. DOI: 10.1007/s00894-017-3208-0. 59

[296] M. Somil Chandrakant Mehta and Alain DuFour. Inhibition of silicascale using amine-terminated polyoxyalkylene, U.S. 2017/0197855A1, n.d. 60

[297] M. Somil Chandrakant Mehta and Alain DuFour. Inhibition of siliscascale using hydrophobcally modified alkylene oxide urethane copolymer, U.S. 2017/0190603 A1, n.d. 60

[298] The Dow Chemical Company, FILMTEC Membranes: Water Chemistry and Pretreatment, pages 1–2, 2006. http://www.filmtec.com 64, 66

[299] R. Sheikholeslami, I. S. Al-Mutaz, T. Koo, and A. Young. Pre-treatment and the effect of cations and anions on prevention of silica fouling. *Desalination*, 139:83–95, 2001. DOI: 10.1016/s0011-9164(01)00297-1. 65, 66

[300] Z. Amjad and R. W. Zuhl. The role of water chemistry on preventing silica fouling in industrial water system, pages 1–14, 2010. 65

[301] I. S. Al-mutaz and I. A. Al-anezi. Silica reduction in reverse osmosis desalting plants. *Engineering*, 2, 2002. 66, 67

[302] I. S. Al-Mutaz and I. A. Al-Anezi. Silica removal during lime softening in water treatment plant. *Int. Conf. Water Resour. Arid Environ.*, pages 1–10, 2004. http://repository.ksu.edu.sa/jspui/handle/123456789/3456. 66

[303] H. Roques. *Chemical Water Treatment: Principles and Practice*, Wiley-VCH, Publ. Inc., New York, 1996. 66

[304] F. K. Lindsay and J. W. Ryznar. Removal of silica from water by sodium aluminate. *Ind. Eng. Chem.*, 31:859–861, 1939. DOI: 10.1021/ie50355a016. 66

[305] A. M. Al-Rehaili. Comparative chemical clarification for silica removal from RO groundwater feed. *Desalination*, 159:21–31, 2003. DOI: 10.1016/s0011-9164(03)90042-7. 66

[306] https://www.finishing.com/429/32.shtml 66, 67

[307] J. N. Apell and T. H. Boyer. Combined ion exchange treatment for removal of dissolved organic matter and hardness. *Water Res.*, 44:2419–2430, 2010. DOI: 10.1016/j.watres.2010.01.004. 68

[308] R. C. Smith and A. K. Sengupta. Integrating tunable anion exchange with reverse osmosis for enhanced recovery during inland brackish water desalination. *Environ. Sci. Technol.*, 49:5637–5644, 2015. DOI: 10.1021/es505439p. 68

[309] J. E. Greenleaf and A. K. Sengupta. Environmentally benign hardness removal using ion exchange fibers and snowmelt. *Environ. Sci. Technol.*, 40:370–376, 2006. DOI: 10.1021/es051702x. 68

[310] B. Van Der Bruggen, K. Everaert, D. Wilms, and C. Vandecasteele. Application of nanofiltration for removal of pesticides, nitrate and hardness from ground water: Rejection properties and economic evaluation. *J. Memb. Sci.*, 193:239–248, 2001. DOI: 10.1016/S0376-7388(01)00517-8. 68

[311] B. Tomaszewska, M. Rajca, E. Kmiecik, M. Bodzek, W. Bujakowski, K. Wator, and M. Tyszer. The influence of selected factors on the effectiveness of pre-treatment of geothermal water during the nanofiltration process. *Desalination*, 406:74–82, 2017. DOI: 10.1016/j.desal.2016.07.007. 68

[312] J. Schaep, B. Van Der Bruggen, S. Uytterhoeven, R. Croux, C. Vandecasteele, D. Wilms, E. Van Houtte, and F. Vanlerberghe. Removal of hardness from groundwater by nanofiltration. *Desalination*, 119:295–301, 1998. DOI: 10.1016/s0011-9164(98)00172-6. 68

[313] E. Kmiecik, M. Bodzek, K. Wator, M. Tyszer, B. Tomaszewska, and M. Rajca. Prevention of scaling during the desalination of geothermal water by means of nanofiltration. *Desalin. Water Treat.*, 73:198–207, 2017. DOI: 10.5004/dwt.2017.20557. 68

[314] M. S. Mohsen, J. O. Jaber, and M. D. Afonso. Desalination of brackish water by nanofiltration and reverse osmosis. *Desalination*, 157:167, 2003. DOI: 10.1016/s0011-9164(03)00397-7. 69

[315] J. Morillo, J. Usero, D. Rosado, H. El, A. Riaza, and F. Bernaola. Comparative study of brine management technologies for desalination plants. *Desalination*, 336:32–49, 2020. DOI: 10.1016/j.desal.2013.12.038. 69, 92, 98, 102

[316] H. Strathmann. Electrodialysis, a mature technology with a multitude of new applications. *Desalination*, 264:268–288, 2010. DOI: 10.1016/j.desal.2010.04.069. 69, 99

[317] N. Kabay, M. Demircioğlu, E. Ersöz, and I. Kurucaovali. Removal of calcium and magnesium hardness of electrodialysis. *Desalination*, 149:343–349, 2002. DOI: 10.1016/s0011-9164(02)00807-x. 69

[318] S. R. Shah, S. L. Walter, and Winter, A. Using feed-forward voltage-control to increase the ion removal rate during batch electrodialysis desalination of brackish water. *Desalination*, 457:62–74, 2019. DOI: 10.1016/j.desal.2019.01.022. 69

[319] Z. Amof, B. Barioub, N. Mameri, M. Taky, S. Nicolasb, and A. Elmidaoui. Desalination fluoride removal corn brackish water by electrodialysis. *Desalination*, 133:215–223, 2001. www.elsevier.com/locate/desal 69

[320] F. Hell, J. Lahnsteiner, H. Frischherz, and G. Baumgartner. Experience with full-scale electrodialysis for nitrate and hardness removal. *Desalination*, 117:173–180, 1998. DOI: 10.1016/s0011-9164(98)00088-5. 69

[321] S. Lahnid, M. Tahaikt, K. Elaroui, I. Idrissi, M. Hafsi, I. Laaziz, Z. Amor, F. Tiyal, and A. Elmidaoui. Economic evaluation of fluoride removal by electrodialysis. *Desalination*, 230:213–219, 2008. DOI: 10.1016/j.desal.2007.11.027. 69

[322] R. K. McGovern, A. M. Weiner, L. Sun, C. G. Chambers, S. M. Zubair, J. H. Lienhard V. On the cost of electrodialysis for the desalination of high salinity feeds. *Appl. Energy*, 136:649–661, 2014. DOI: 10.1016/j.apenergy.2014.09.050. 69

[323] R. K. McGovern, S. M. Zubair, and J. H. Lienhard V. The cost effectiveness of electrodialysis for diverse salinity applications. *Desalination*, 348:57–65, 2014. DOI: 10.1016/j.desal.2014.06.010. 69

[324] M. Demircioğlu, N. Kabay, E. Ersöz, I. Kurucaovali, Ç. Şafak, and N. Gizli. Cost comparison and efficiency modeling in the electrodialysis of brine. *Desalination*, 136:317–323, 2001. DOI: 10.1016/s0011-9164(01)00194-1. 69

[325] H. J. Lee, F. Sarfert, H. Strathmann, and S. H. Moon. Designing of an electrodialysis desalination plant. *Desalination*, 142:267–286, 2002. DOI: 10.1016/s0011-9164(02)00208-4. 69

[326] C. Casademont, G. Pourcelly, and L. Bazinet. Effect of magnesium/calcium ratios in solutions treated by electrodialysis: Morphological characterization and identification of anion exchange membrane fouling. *J. Colloid Interface Sci.*, 322:215–223, 2008. DOI: 10.1016/j.jcis.2008.02.068. 69

[327] A. Al-Karaghouli and L. L. Kazmerski. Energy consumption and water production cost of conventional and renewable-energy-powered desalination processes. *Renew. Sustain. Energy Rev.*, 24:343–356, 2013. DOI: 10.1016/j.rser.2012.12.064. 70, 88, 89, 90, 98, 99, 131

[328] H. J. Ahn, J. H. Lee, Y. Jeong, J. H. Lee, C. S. Chi, and H. J. Oh. Nanostructured carbon cloth electrode for desalination from aqueous solutions. *Mater. Sci. Eng. A*, pages 448–451 and pages 841–845, 2007. DOI: 10.1016/j.msea.2006.02.448. 70

[329] E. Avraham, B. Yaniv, A. Soffer, and D. Aurbach. Developing ion electroadsorption stereoselectivity, by pore size adjustment with chemical vapor deposition onto active carbon fiber electrodes. Case of Ca^{2+}/Na^{+} separation in water capacitive desalination. *J. Phys. Chem. C*, 112:7385–7389, 2008. DOI: 10.1021/jp711706z. 70

[330] T. J. Welgemoed and C. F. Schutte. Capacitive deionization technology™: An alternative desalination solution. *Desalination*, 183:327–340, 2005. DOI: 10.1016/j.desal.2005.02.054. 70

[331] Y. Oren. Capacitive deionization (CDI) for desalination and water treatment— past, present, and future (a review). *Desalination*, 228:10–29, 2008. DOI: 10.1016/j.desal.2007.08.005. 70

[332] S. Porada, R. Zhao, A. Van Der Wal, V. Presser, and P. M. Biesheuvel. Review on the science and technology of water desalination by capacitive deionization. *Prog. Mater. Sci.*, 58:1388–1442, 2013. DOI: 10.1016/j.pmatsci.2013.03.005. 70

[333] N. Spoljaric and W. A. Crawford. Removal of contaminants from landfill leachates by filtration through glauconitic greensands. *Environ. Geol.*, 2:359–363, 1979. DOI: 10.1007/bf02380510. 71

[334] N. Spoljaric and W. A. Crawford. Glauconitic greensand: A possible filter of heavy metal cations from polluted waters. *Environ. Geol.*, 2:215–221, 1978. DOI: 10.1007/bf02380487. 71

[335] O. J. Haoa and C. M. Tsai. The removal of metal and ammonium by natural glauconite, 13:203–212, 1987. DOI: 10.1016/0160-4120(87)90091-2. 71

[336] M. E. Doumer, M. Vidal, A. S. Mangrich, and A. Rigol. Feasibility of using low-cost, by product materials as sorbents to remove heavy metals from aqueous solutions. *Environ. Technol.*, 0:1–10, UK, 2018. DOI: 10.1080/09593330.2018.1440011. 71

[337] M. J. Tallman, C. Santner, and R. B. Miller. Materials for removing toxic metals from wastewater, 1, 2006. DOI: 10.1038/incomms1464. 71

[338] N. Khatri, S. Tyagi, and D. Rawtani. Recent strategies for the removal of iron from water: A review. *J. Water Process Eng.*, 19:291–304, 2017. DOI: 10.1016/j.jwpe.2017.08.015. 71

[339] J. Wu, E. De Antonio Mario, B. Yang, C. Liu, F. Jia, and S. Song. Efficient removal of $Hg^{2}+$ in aqueous solution with fishbone charcoal as adsorbent. *Environ. Sci. Pollut. Res.*, 2018. DOI: 10.1007/s11356-017-1007-x. 71

[340] M. Ohno, C. Manalo, L. Rossetto, T. Okuda, S. Nakai, and W. Nishijima. Effect of coexisting metal ions on the degradation of polyamide reverse osmosis membrane by hypochlorite treatment. *Desalination*, 381:126–134, 2016. DOI: 10.1016/j.desal.2015.12.005. 71, 72

[341] L. M. Sayre. Metal ion catalysis of amide hydrolysis. *J. Am. Chem. Soc.*, 108:1632–1635, 1986. DOI: 10.1021/ja00267a037. 72

[342] J. S. Gill. Inhibition of silica—silicate deposit in industrial waters. *Colloids Surfaces A Physicochem. Eng. Asp.*, 74:101–106, 1993. DOI: 10.1016/0927-7757(93)80401-y. 72

[343] M. Kennedy, A. Jaljuli, G.-J. Witkamp, I. Bremere, J. Schippers, and S. Mhyio. Prevention of silica scale in membrane systems: Removal of monomer and polymer silica. *Desalination*, 132:89–100, 2003. DOI: 10.1016/s0011-9164(00)00138-7. 72

[344] H. Sugita, K. Kato, A. Ueda, I. Matsunaga, Y. Sakurai, K. Yasuda, Y. Bando, and M. Nakamura. Field tests on silica removal from geothermal brines in Sumikawa and Onuma geothermal areas. *J. Chem. Eng. Japan.*, 32:696–700, 1999. DOI: 10.1252/jcej.32.696. 72

[345] E. G. Darton. RO plant experiences with high silica waters in the Canary Islands. *Desalination*, 124:33–41, 1999. DOI: 10.1016/s0011-9164(99)00086-7. 72

[346] H. Sugita, Y. Bando, and M. Nakamura. Removal of silica from geothermal brine by seeding method using silica gel. *J. Chem. Eng. Japan.*, 31:150–152, 1998. DOI: 10.1252/jcej.31.150. 72

[347] S. F. Roca and R. Cited. Method and apparatus for high efficiency reverse osmosis operation, 2 2006. 72

[348] J. Thomson, M. Slunjski, M. Fabig, and G. Crisp. Potable water HERO process at Yalgoo. *Water Pract. Technol.*, 4, 2009. DOI: 10.2166/wpt.2009.033. 72, 73

[349] Y. Chen, J. C. Baygents, and J. Farrell. Evaluating electrocoagulation and chemical coagulation for removing dissolved silica from high efficiency reverse osmosis (HERO) concentrate solutions. *J. Water Process Eng.*, 16:50–55, 2017. DOI: 10.1016/j.jwpe.2016.12.008. 72

[350] D. Mukhopadhyay. Method for high efficiency reverse osmosis operation, U.S. Patent Application no. 14/144, 423, 2013.

[351] J. Kucera. *Reverse Osmosis: Industrial Processes and Applications*, John Wiley & Sons, 2015. DOI: 10.1002/9781119145776. 72

[352] T. Nguyen, F. A. Roddick, and L. Fan. Biofouling of water treatment membranes: A review of the underlying causes, monitoring techniques and control measures. *Membranes (Basel)*, 2:804–840, 2012. DOI: 10.3390/membranes2040804. 73, 74

[353] D. Mukhopadhyay. High purity water produced by reverse osmosis, U.S. 8,758,720 B2, 2014. https://patentimages.storage.googleapis.com/77/25/61/4462649cef8c07/US8758720.pdf 73

[354] OPUS®, Optimized Pretreatment and Unique Separation Technology for Surface Water Discharge or Reuse, (n.d.). 73, 74

[355] OPUS, (n.d.). http://technomaps.veoliawatertechnologies.com/opus/en/ 73

[356] Veolia Water Technologies, OPUSII, (n.d.). http://technomaps.veoliawatertechnologies.com/processes/lib/pdfs/productbrochures/key_technologies/2471,OPUSII_brochure.pdf 75

[357] T. Client and T. Solution. Oil & Gas case study OPUS®technology treats produced water for aquifer recharge. The client's needs case study scope of work and technologies, 2008. 75, 76

[358] C. J. Gabelich, A. Rahardianto, C. R. Northrup, T. I. Yun, and Y. Cohen. Process evaluation of intermediate chemical demineralization for water recovery enhancement in production-scale brackish water desalting. *Desalination*, 272:36–45, 2011. DOI: 10.1016/j.desal.2010.12.050.

[359] Y. Cohen, M. D. Williams, J. Gao, A. Rahardianto, and C. J. Gabelich. High recovery membrane desalting of low-salinity brackish water: Integration of accelerated precipitation softening with membrane RO. *J. Memb. Sci.*, 289:123–137, 2006. DOI: 10.1016/j.memsci.2006.11.043. 76

[360] C. J. Gabelich, M. D. Williams, A. Rahardianto, J. C. Franklin, and Y. Cohen. High-recovery reverse osmosis desalination using intermediate chemical demineralization. *J. Memb. Sci.*, 301:131–141, 2007. DOI: 10.1016/j.memsci.2007.06.007. 76

[361] A. Zhu, P. D. Christofides, and Y. Cohen. On RO membrane and energy costs and associated incentives for future enhancements of membrane permeability. *J. Memb. Sci.*, 344:1–5, 2009. DOI: 10.1016/j.memsci.2009.08.006. 76

[362] A. Drak. MaxH2O DESALTER technology treats calcium sulphate saturated wastewater, a perfect solution for treating acid mine drainage wastewater. *Int. Water Conf.*, 2018. 77, 78

[363] S. Wait. High recovery RO with pellet reactor: Case study review. *AMTA-SWMOA Technol. Transf. Work. Enhanced Recover. NF/RO Syst.*, Albuquerque, NM, 2019. 77, 78

[364] M. Uchymiak, H. Gu, Y. Cohen, A. R. Bartman, and P. D. Christofides. Self-adaptive feed flow reversal operation of reverse osmosis desalination. *Desalination*, 308:63–72, 2012. DOI: 10.1016/j.desal.2012.07.041. 78

[365] M. Uchymiak, A. R. Bartman, N. Daltrophe, M. Weissman, J. Gilron, P. D. Christofides, W. J. Kaiser, and Y. Cohen. Brackish water reverse osmosis (BWRO) operation in feed flow reversal mode using an ex situ scale observation detector (EXSOD). *J. Memb. Sci.*, 341:60–66, 2009. DOI: 10.1016/j.memsci.2009.05.039.

[366] A. R. Bartman, C. W. McFall, P. D. Christofides, and Y. Cohen. Model predictive control of feed flow reversal in a reverse osmosis desalination process. *Proc. Am. Control Conf.*, 19:4860–4867, 2009. DOI: 10.1109/acc.2009.5160150. 80

[367] R. Erlitzki. 75% is totally 70s: New water, high recovery, concentrate management. *AMTA-SWMOA Technol. Transf. Work. Enhanced Recover. NF/RO Syst.*, Albuquerque, NM, 2019. 79

[368] J. R. Werber, A. Deshmukh, and M. Elimelech. Can batch or semi-batch processes save energy in reverse-osmosis desalination? *Desalination*, 402:109–122, 2017. DOI: 10.1016/j.desal.2016.09.028. 80

[369] S. Mirza. Reduction of energy consumption in process plants using nanofiltration and reverse osmosis. *Desalination*, 224:132–142, 2008. DOI: 10.1016/j.desal.2007.04.084. 80

[370] A. J. Tarquin, M. P. Fahy, and J. E. Balliew. Concentrate volume reduction research. pages 3507–3518, El Paso, TX, 2010. DOI: 10.1061/41114(371)358. 80

[371] A. Tarquin and G. Delgado. Concentrate enhanced recovery reverse osmosis: A new process for RO concentrate and brackish water treatment. *Am. Inst. Chem. Eng. Meet.*, Pittsburg, PA, October, 2012. 80

[372] R. Stover and G. Carpenter. Creative solutions and innovative strategies for today's water challenges: Blue paper, 2017. 80

[373] R. L. Stover. Permeate recovery and flux maximization in semibatch reverse osmosis. *IDA J. Desalin. Water Reuse*, 5:10–14, 2013. DOI: 10.1179/2051645213y.0000000002. 80

[374] Desalitech, Industrial Closed Circuit Reverse Osmosis Systems, (n.d.). https://www.desalitech.com/industrial-closed-circuit-reverse-osmosis-systems/ 80

[375] M. Boyd. What are practical approaches to improve RO technology? *AWT Annu. Conf.*, Orlando, FL, 2018. 80, 81, 83, 84

[376] M. Boyd. Closed circuit reverse osmosis (CCRO). *AMTA-SWMOA Technol. Transf. Work. Enhanced Recover. NF/RO Syst.*, Albuquerque, NM, 2019. 81

[377] D. M. Warsinger, E. W. Tow, K. G. Nayar, L. A. Maswadeh, and J. H. Lienhard V. Energy efficiency of batch and semi-batch (CCRO) reverse osmosis desalination. *Water Res.*, 106:272–282, 2016. DOI: 10.1016/j.watres.2016.09.029. 82

[378] C. Bertoia. PFAS removal for drinking water and discharge: Protecting what matters most. *Am. Water Work. Assoc.*, Webinar, 2018. 83

[379] B. Liberman. Pulse flow RO—concentrate squeezer. *11th IWA Int. Conf. Water Reclam. Reuse*, 2017. 83, 84, 85

[380] IDE PFRO, (n.d.). https://www.ide-tech.com/en/solutions/desalination/brineminimization/?data=item_1

[381] MAXH2O, (n.d.). https://www.ide-tech.com/en/solutions/desalination/brineminimization/?data=item_1 85

[382] A. Subramani and J. G. Jacangelo. Treatment technologies for reverse osmosis concentrate volume minimization: A review. *Sep. Purif. Technol.*, 122:472–489, 2014. DOI: 10.1016/j.seppur.2013.12.004.

[383] M. A. Cappelle. High recovery inland desalination: A technical and economic performance evaluation of zero discharge desalination, 2015.

[384] A. Subramani and J. G. Jacangelo. Emerging desalination technologies for water treatment: A critical review. *Water Res.*, 75:164–187, 2015. DOI: 10.1016/j.watres.2015.02.032. 87, 115

[385] B. K. Pramanik, L. Shu, and V. Jegatheesan. A review of the management and treatment of brine solutions. *Environ. Sci. Water Res. Technol.*, 3:625–658, 2017. DOI: 10.1039/c6ew00339g. 87

[386] E. Drioli, N. Geisma, L. Katzir, J. Gilron, F. Macedonio, and S. Simone. Wind-aided intensified evaporation (WAIV) and membrane crystallizer (MCr) integrated brackish water desalination process: Advantages and drawbacks. *Desalination*, 273:127–135, 2011. DOI: 10.1016/j.desal.2010.12.002. 89

[387] L. Katzir, E. Korngold, N. Daltrophe, J. Gilron, Y. Oren, R. Mesalem, and Y. Volkmann. WAIV—wind aided intensified evaporation for brine volume reduction and generating mineral by products. *Desalin. Water Treat.*, 13:63–73, 2010. DOI: 10.5004/dwt.2010.772. 90, 91

[388] P. Chelme-Ayala, D. W. Smith, and M. G. El-Din. Membrane concentrate management options: A comprehensive critical review 1, 1119:1107–1119, 2009. DOI: 10.1139/l09-042. 90, 92

[389] D. Soliz, E. P. Glenn, R. Seaman, M. Yoklic, S. G. Nelson, and P. Brown. Water consumption, irrigation efficiency and nutritional value of Atriplex lentiformis grown on reverse osmosis brine in a desert irrigation district. *Agric. Ecosyst. Environ.*, 140:473–483, 2011. DOI: 10.1016/j.agee.2011.01.012. 92

[390] T. Y. Cath, A. E. Childress, and M. Elimelech. Forward osmosis: Principles, applications, and recent developments. *J. Memb. Sci.*, 281:70–87, 2006. DOI: 10.1016/j.memsci.2006.05.048. 92

[391] Y. Oh, S. Lee, M. Elimelech, S. Lee, and S. Hong. Effect of hydraulic pressure and membrane orientation on water flux and reverse solute flux in pressure assisted osmosis. *J. Memb. Sci.*, 465:159–166, 2014. DOI: 10.1016/j.memsci.2014.04.008. 92

[392] A. Achilli and A. E. Childress. Pressure retarded osmosis: From the vision of Sidney Loeb to the first prototype installation. *Desalination*, 261:205–211, 2010. DOI: 10.1016/j.desal.2010.06.017. 92

[393] T. V. Bartholomew, M. S. Mauter, L. Mey, J. T. Arena, and N. S. Siefert. Osmotically assisted reverse osmosis for high salinity brine treatment. *Desalination*, 421:3–11, 2017. DOI: 10.1016/j.desal.2017.04.012. 92, 93, 94

[394] A. T. Bouma and J. H. Lienhard V. Split-feed counter flow reverse osmosis for brine concentration. *Desalination*, 445:280–291, 2018. DOI: 10.1016/j.desal.2018.07.011. 94, 95, 96

[395] I. Ozturk, K. Roest, H. Ozgun, J. B. van Lier, M. E. Ersahin, and R. K. Dereli. A review on dynamic membrane filtration: Materials, applications and future perspectives. *Bioresour. Technol.*, 122:196–206, 2012. DOI: 10.1016/j.biortech.2012.03.086. 95, 97

[396] X. Zhang, Z. Wang, Z. Wu, F. Lu, J. Tong, and L. Zang. Formation of dynamic membrane in an anaerobic membrane bioreactor for municipal wastewater treatment. *Chem. Eng. J.*, 165:175–183, 2010. DOI: 10.1016/j.cej.2010.09.013. 95

[397] Osmoflo Brine Squeezer (OBS) Process, (n.d.). https://www.osmoflo.com/globalassets/ourcapabilities/osmoflo-brine-squeezer---obs.pdf 95, 96

[398] H. Le. Brine squeezer. *AMTA-SWMOA Technol. Transf. Work. Recover. NF/RO Syst.*, Albuquerque, NM, 2019. 95, 97

[399] L. Nickels. Osmoflo develops RO brine squeezer, 2014. https://www.filtsep.com/desalination/news/osmoflo-develops-ro-brine-squeezer/ 95

[400] Images de J. D. Wall P.E. February, AquaSel™Brine Concentration Pilot Project Update, 2017. 98

[401] E. Korngold, L. Aronov, and N. Daltrophe. Electrodialysis of brine solutions discharged from an RO plant. *Desalination*, 242:215–227, 2009. DOI: 10.1016/j.desal.2008.04.008. 98

[402] J. D. Wall. AquaSel™Brine Concentration Pilot Project Update, 2017. 98

[403] T. A. Davis. Centre for membrane science and technology seminar, 2014. 99, 100

[404] R. Bond, B. Batchelor, T. Davis, and B. Klayman. Zero liquid discharge desalination of brackish water with an innovative form of electrodialysis: Electrodialysis metathesis, pages 38–42, 2011. 99, 100, 101

[405] R. Bond. Demonstration of a new electrodialysis technology to reduce energy required for salinity management: Final project report, California Energy Commission, Energy Research and Development Division, 2015. 99

[406] T. A. Davis. Demonstration of zero discharge desalination (ZDD), 2014. 100, 101

[407] J. Morillo, J. Usero, D. Rosado, H. El Bakouri, A. Riaza, and F.-J. Bernaola. Comparative study of brine management technologies for desalination plants. *Desalination*, 336:32–49, 2014. DOI: 10.1016/j.desal.2013.12.038. 101, 120, 122

[408] S. Ahuja, M. C. Larsen, J. L. Eimers, C. L. Patterson, S. Sengupta, and J. L. Schnoor. *Comprehensive Water Quality and Purification*, Elsevier Amsterdam, 2014. 101, 106, 107, 109

[409] D. González, J. Amigo, and F. Suárez. Membrane distillation: Perspectives for sustainable and improved desalination. *Renew. Sustain. Energy Rev.*, 80:238–259, 2017. DOI: 10.1016/j.rser.2017.05.078. 102, 140, 141, 142, 143

[410] C. R. Martinetti, A. E. Childress, and T. Y. Cath. High recovery of concentrated RO brines using forward osmosis and membrane distillation. *J. Memb. Sci.*, 331:31–39, 2009. DOI: 10.1016/j.memsci.2009.01.003. 103, 105, 108, 118, 119

[411] S. Lee, Y. Kim, A. S. Kim, S. Hong, S. Lee, Y. Kim, A. S. Kim, and S. Hong. Evaluation of membrane-based desalting processes for RO brine treatment. *Desalin. Water Treat.*, 3994:0, 2017. DOI: 10.1080/19443994.2015.1030120. 103, 104, 105

[412] D. Winter, J. Koschikowski, and M. Wieghaus. Desalination using membrane distillation: Experimental studies on full scale spiral wound modules. *J. Memb. Sci.*, 375:104–112, 2011. DOI: 10.1016/j.memsci.2011.03.030. 104, 105

[413] Y. Ghalavand, M. S. Hatamipour, and A. Rahimi. A review on energy consumption of desalination processes. *Desalin. Water Treat.*, 3994:1–16, 2017. DOI: 10.1080/19443994.2014.892837. 105

[414] P. Xu, T. Y. Cath, A. P. Robertson, M. Reinhard, J. O. Leckie, and J. E. Drewes. Critical review of desalination concentrate management, treatment and beneficial use. *Environ. Eng. Sci.*, 30:502–514, 2013. DOI: 10.1089/ees.2012.0348. 106, 107, 108, 115, 116, 117, 118

[415] Dewvaporation Desalination 5 , 000-Gallon-Per-Day Pilot Plant, 2008. 106, 107

[416] Altela Inc., RO Wastewater Concentrate Demonstration Goals, 2015. 106, 107

[417] L. M. Vane. Water recovery from brines and salt-saturated solutions: Operability and thermodynamic efficiency considerations for desalination technologies. *J. Chem. Technol. Biotechnol.*, 92:2506–2518, 2017. DOI: 10.1002/jctb.5225. 108

[418] D. L. Shaffer, J. R. Werber, H. Jaramillo, S. Lin, and M. Elimelech. Forward osmosis: Where are we now? *Desalination*, 356:271–284, 2015. DOI: 10.1016/j.desal.2014.10.031. 108, 109

[419] R. L. McGinnis, N. T. Hancock, M. S. Nowosielski-Slepowron, and G. D. McGurgan. Pilot demonstration of the NH3/CO2 forward osmosis desalination process on high salinity brines. *Desalination*, 312:67–74, 2013. DOI: 10.1016/j.desal.2012.11.032. 109, 110

[420] T. Tong and M. Elimelech. The global rise of zero liquid discharge for wastewater management: Drivers, technologies, and future directions. *Environ. Sci. Technol.*, 2016. DOI: 10.1021/acs.est.6b01000. 110, 113, 114, 115

[421] Evaporation systems for water desalination, (n.d.). https://blogen.condorchem.com/evaporation-systems-water-desalination/#.XKFa35hKiUl 113

[422] Y. Oren, E. Korngold, N. Daltrophe, R. Messalem, Y. Volkman, L. Aronov, M. Weismann, N. Bouriakov, P. Glueckstern, and J. Gilron. Pilot studies on high recovery BWRO-EDR for near zero liquid discharge approach. *Desalination*, 261:321–330, 2010. DOI: 10.1016/j.desal.2010.06.010. 114

[423] K. Loganathan, P. Chelme-Ayala, and M. G. El-Din. Treatment of basal water using a hybrid electrodialysis reversal—reverse osmosis system combined with a low-temperature crystallizer for near-zero liquid discharge. *Desalination*, 363:92–98, 2015. DOI: 10.1016/j.desal.2015.01.020. 114

[424] R. Ashu, E. Curcio, E. Brauns, W. Van Baak, E. Fontananova, and G. Di. Membrane distillation and reverse electrodialysis for near-zero liquid discharge and low energy seawater desalination, 496:325–333, 2015. DOI: 10.1016/j.memsci.2015.09.008. 115

[425] J. A. Ruskowitz, F. Suárez, S. W. Tyler, and A. E. Childress. Evaporation suppression and solar energy collection in a salt-gradient solar pond. *Sol. Energy*, 99:36–46, 2014. DOI: 10.1016/j.solener.2013.10.035. 116

[426] H. Lu, A. H. P. Swift, H. D. Hein, and J. C. Walton. Advancements in salinity gradient solar pond technology based on sixteen years of operational experience. *J. Sol. Energy Eng.*, 126:759–767, 2004. DOI: 10.1115/1.1667977. 116

[427] B. E. Logan and M. Elimelech. Membrane-based processes for sustainable power generation using water. *Nature*, 488:313, 2012. DOI: 10.1038/nature11477. 117

[428] F. La Mantia, M. Pasta, H. D. Deshazer, B. E. Logan, and Y. Cui. Batteries for efficient energy extraction from a water salinity difference. *Nano Lett.*, 11:1810–1813, 2011. DOI: 10.1021/nl200500s. 117, 118

[429] A. Ravizky and N. Nadav. Salt production by the evaporation of SWRO brine in Eilat: A success story. *Desalination*, 205:374–379, 2007. DOI: 10.1016/j.desal.2006.03.559. 118

[430] EWM, Full Recovery Desalination, (n.d.). https://envirowaterminerals.com/index.html 119

[431] T. J. Rittof. Selective salt recovery from mixed salt brine, U.S. 8,603,192 B2, 2013. 119

[432] M. Badruzzaman, J. Oppenheimer, S. Adham, and M. Kumar. Innovative beneficial reuse of reverse osmosis concentrate using bipolar membrane electrodialysis and electrochlorination processes, 326:392–399, 2009. DOI: 10.1016/j.memsci.2008.10.018. 119

[433] J. R. Davis, Y. Chen, J. C. Baygents, and J. Farrell. Production of acids and bases for ion exchange regeneration from dilute salt solutions using bipolar membrane electrodialysis, pages 3–8, 2015. DOI: 10.1021/acssuschemeng.5b00654. 120, 121

[434] Y. Yang, X. Gao, A. Fan, L. Fu, and C. Gao. An innovative beneficial reuse of seawater concentrate using bipolar membrane electrodialysis. *J. Memb. Sci.*, 449:119–126, 2014. DOI: 10.1016/j.memsci.2013.07.066. 121, 122

[435] C. Fernandez-Gonzalez, A. Dominguez-Ramos, R. Ibañez, and A. Irabien. Electrodialysis with bipolar membranes for valorization of brines electrodialysis with bipolar membranes for valorization of brines, 2119, 2016. DOI: 10.1080/15422119.2015.1128951. 122, 124

[436] Y. Zhao. A novel technology of carbon dioxide adsorption and mineralization via seawater decalcification by bipolar membrane electrodialysis system with a crystallizer. *Chem. Eng. J.*, (n.d.). DOI: 10.1016/j.cej.2019.122542. 124

[437] S. Arabia. Water desalination systems powered by renewable energy sources: Review. *Renew. Sustain. Energy Rev.*, 16:1537–1556, 2012. DOI: 10.1016/j.rser.2011.11.002. 125, 130

[438] S. Li, Y. Cai, A. I. Schäfer, and B. S. Richards. Renewable energy powered membrane technology: A review of the reliability of photovoltaic-powered membrane system components for brackish water desalination. *Appl. Energy*, 253:113524, 2019. DOI: 10.1016/j.apenergy.2019.113524. 125

[439] A. Kasaeian, F. Rajaee, and W. Yan. Osmotic desalination by solar energy: A critical review. *Renew. Energy*, 2018. DOI: 10.1016/j.renene.2018.09.038. 125

[440] A. Ali, R. A. Tufa, F. Macedonio, E. Curcio, and E. Drioli. Membrane technology in renewable-energy-driven desalination. *Renew. Sustain. Energy Rev.*, 81:1–21, 2018. DOI: 10.1016/j.rser.2017.07.047. 125, 126, 127, 128, 129, 130, 131, 132, 133, 134, 138, 140, 141, 142, 144

[441] F. E. Ahmed, R. Hashaikeh, and N. Hilal. Solar powered desalination—Technology, energy and future outlook. *Desalination*, 453:54–76, 2019. DOI: 10.1016/j.desal.2018.12.002. 125, 127

[442] S. A. Kalogirou. Seawater desalination using renewable energy sources. *Prog. Energy Combust. Sci.*, 31:242–281, 2005. DOI: 10.1016/j.pecs.2005.03.001. 126

[443] H. Mahmoudi, N. Spahis, M. F. Goosen, N. Ghaffour, N. Drouiche, and A. Ouagued. Application of geothermal energy for heating and fresh water production in a brackish water greenhouse desalination unit: A case study from Algeria. *Renew. Sustain. Energy Rev.*, 14:512–517, 2010. DOI: 10.1016/j.rser.2009.07.038. 126

[444] A. Falcão. Wave energy utilization: A review of the technologies renewable and sustainable energy. *Sci. Direct.*, pages 1–11, 2009. DOI: 10.1016/j.rser.2009.11.003. 128

[445] T. He and L. Yan. Application of alternative energy integration technology in seawater desalination. *Desalination*, 249:104–108, 2009. DOI: 10.1016/j.desal.2008.07.026. 128

[446] L. Scheinman. *The International Atomic Energy Agency and World Nuclear Order*. Routledge, 2016. DOI: 10.4324/9781315667676. 128

[447] I. G. Sánchez-Cervera, K. C. Kavvadias, and I. Khamis. DE-TOP: A new IAEA tool for the thermodynamic evaluation of nuclear desalination. *Desalination*, 321:103–109, 2013. DOI: 10.1016/j.desal.2011.10.005. 128

[448] A. Adamantiades and I. Kessides. Nuclear power for sustainable development: Current status and future prospects. *Energy Policy*, 37:5149–5166, 2009. DOI: 10.1016/j.enpol.2009.07.052. 128

[449] R. Tarjanne and A. Kivistö. Comparison of electricity generation costs. *Res. Report/Faculty Technol. Dep. Energy Environ. Technol.*, 2008. 128

[450] A. Al-Othman, N. N. Darwish, M. Qasim, M. Tawalbeh, N. A. Darwish, and N. Hilal. Nuclear desalination: A state-of-the-art review. *Desalination*, 457:39–61, 2019. DOI: 10.1016/j.desal.2019.01.002. 128, 129, 134, 135

[451] J. Veerman, M. Saakes, S. J. Metz, and G. J. Harmsen. Electrical power from sea and river water by reverse electrodialysis: A first step from the laboratory to a real power plant. *Environ. Sci. Technol.*, 44:9207–9212, 2010. DOI: 10.1021/es1009345. 129

[452] J. Veerman, J. W. Post, M. Saakes, S. J. Metz, and G. J. Harmsen. Reducing power losses caused by ionic shortcut currents in reverse electrodialysis stacks by a validated model. *J. Memb. Sci.*, 310:418–430, 2008. DOI: 10.1016/j.memsci.2007.11.032. 129

[453] J. W. Post, H. V. M. Hamelers, and C. J. N. Buisman. Energy recovery from controlled mixing salt and fresh water with a reverse electrodialysis system. *Environ. Sci. Technol.*, 42:5785–5790, 2008. DOI: 10.1021/es8004317. 129

[454] B. B. Sales, M. Saakes, J. W. Post, C. J. N. Buisman, P. M. Biesheuvel, and H. V. M. Hamelers. Direct power production from a water salinity difference in a membrane-modified supercapacitor flow cell. *Environ. Sci. Technol.*, 44:5661–5665, 2010. DOI: 10.1021/es100852a. 129

[455] Z. Jia, B. Wang, S. Song, and Y. Fan. Blue energy: Current technologies for sustainable power generation from water salinity gradient. *Renew. Sustain. Energy Rev.*, 31:91–100, 2014. DOI: 10.1016/j.rser.2013.11.049. 129

[456] M. T. Mito, X. Ma, H. Albu, and P. A. Davies. Reverse osmosis (RO) membrane desalination driven by wind and solar photovoltaic (PV) energy: State-of-the-art and challenges for large scale implementation. *Renew. Sustain. Energy Rev.*, 112:669–685, 2019. DOI: 10.1016/j.rser.2019.06.008. 130, 131, 132, 135, 136, 137, 138

[457] M. Thomson and D. Infield. Laboratory demonstration of a photovoltaic-powered seawater reverse-osmosis system without batteries. *Desalination*, 183:105–111, 2005. DOI: 10.1016/j.desal.2005.03.031. 130

[458] A. M. Helal, S. A. Al-Malek, and E. S. Al-Katheeri. Economic feasibility of alternative designs of a PV-RO desalination unit for remote areas in the United Arab Emirates. *Desalination*, 221:1–16, 2008. DOI: 10.1016/j.desal.2007.01.064. 131, 132

[459] E. S. Mohamed, G. Papadakis, E. Mathioulakis, and V. Belessiotis. A direct coupled photovoltaic seawater reverse osmosis desalination system toward battery based systems—a technical and economical experimental comparative study. *Desalination*, 221:17–22, 2008. DOI: 10.1016/j.desal.2007.01.065. 131

[460] A. Soric, R. Cesaro, P. Perez, E. Guiol, and P. Moulin. Eausmose project desalination by reverse osmosis and batteryless solar energy: Design for a 1 m^3 per day delivery. *Desalination*, 301:67–74, 2012. DOI: 10.1016/j.desal.2012.06.013. 132

[461] B. Peñate and L. García-Rodríguez. Seawater reverse osmosis desalination driven by a solar organic rankine cycle: Design and technology assessment for medium capacity range. *Desalination*, 284:86–91, 2012. DOI: 10.1016/j.desal.2011.08.040. 132

[462] D. Manolakos, E. S. Mohamed, I. Karagiannis, and G. Papadakis. Technical and economic comparison between PV-RO system and RO-solar rankine system. case study: Thirasia Island. *Desalination*, 221:37–46, 2008. DOI: 10.1016/j.desal.2007.01.066.

[463] U. Caldera, D. Bogdanov, and C. Breyer. Local cost of seawater RO desalination based on solar PV and wind energy: A global estimate. *Desalination*, 385:207–216, 2016. DOI: 10.1016/j.desal.2016.02.004. 132

[464] M. Gökçek and Ö. B. Gökçek. Technical and economic evaluation of freshwater production from a wind-powered small-scale seawater reverse osmosis system (WP-SWRO). *Desalination*, 381:47–57, 2016. DOI: 10.1016/j.desal.2015.12.004. 132

[465] W. Lai, Q. Ma, H. Lu, S. Weng, J. Fan, and H. Fang. Effects of wind intermittence and fluctuation on reverse osmosis desalination process and solution strategies. *Desalination*, 395:17–27, 2016. DOI: 10.1016/j.desal.2016.05.019. 132

[466] Y. Kumar, J. Ringenberg, S. S. Depuru, V. K. Devabhaktuni, J. W. Lee, E. Nikolaidis, B. Andersen, and A. Afjeh. Wind energy: Trends and enabling technologies. *Renew. Sustain. Energy Rev.*, 53:209–224, 2016. DOI: 10.1016/j.rser.2015.07.200. 132

[467] S. Loutatidou, N. Liosis, R. Pohl, T. B. M. J. Ouarda, and H. A. Arafat. Wind-powered desalination for strategic water storage: Techno-economic assessment of concept. *Desalination*, 408:36–51, 2017. DOI: 10.1016/j.desal.2017.01.002. 132

[468] E. Tzen, D. Theofilloyianakos, and Z. Kologios. Autonomous reverse osmosis units driven by RE sources experiences and lessons learned. *Desalination*, 221:29–36, 2008. DOI: 10.1016/j.desal.2007.02.048. 132

[469] M. S. Miranda and D. Infield. A wind-powered seawater reverse-osmosis system without batteries. *Desalination*, 153:9–16, 2003. DOI: 10.1016/s0011-9164(02)01088-3. 132

[470] M. A. Abdelkareem, M. E. H. Assad, E. T. Sayed, and B. Soudan. Recent progress in the use of renewable energy sources to power water desalination plants. *Desalination*, 435:97–113, 2018. DOI: 10.1016/j.desal.2017.11.018. 133

[471] P. A. Davies. Wave-powered desalination: Resource assessment and review of technology. *Desalination*, 186:97–109, 2005. DOI: 10.1016/j.desal.2005.03.093. 133

[472] N. Sharmila, P. Jalihal, A. K. Swamy, and M. Ravindran. Wave powered desalination system. *Energy*, 29:1659–1672, 2004. DOI: 10.1016/j.energy.2004.03.099. 133

[473] M. Folley, B. P. Suarez, and T. Whittaker. An autonomous wave-powered desalination system. *Desalination*, 220:412–421, 2008. DOI: 10.1016/j.desal.2007.01.044. 133

[474] M. G. Abdoelatef, R. M. Field, and Y.-K. Lee. Thermodynamic evaluation of coupling APR1400 with a thermal desalination plant. *Int. J. Chem. Mol. Nucl. Mater. Metall. Eng.*, 9:1217–1225, 2015. 134

[475] J. A. Carta, J. Gonzalez, and V. Subiela. Operational analysis of an innovative wind powered reverse osmosis system installed in the Canary Islands. *Sol. Energy*, 75:153–168, 2003. DOI: 10.1016/s0038-092x(03)00247-0.

[476] F. Moreno and A. Pinilla. Preliminary experimental study of a small reverse osmosis wind-powered desalination plant. *Desalination*, 171:257–265, 2005. DOI: 10.1016/j.desal.2004.06.191.

[477] R. Pohl, M. Kaltschmitt, and R. Holländer. Investigation of different operational strategies for the variable operation of a simple reverse osmosis unit. *Desalination*, 249:1280–1287, 2009. DOI: 10.1016/j.desal.2009.06.029.

[478] B. Peñate, F. Castellano, A. Bello, and L. García-Rodríguez. Assessment of a stand-alone gradual capacity reverse osmosis desalination plant to adapt to wind power availability: A case study. *Energy*, 36:4372–4384, 2011. DOI: 10.1016/j.energy.2011.04.005.

[479] K. Bognar, R. Pohl, and F. Behrendt. Seawater reverse osmosis (SWRO) as deferrable load in micro grids. *Desalin. Water Treat.*, 51:1190–1199, 2013. DOI: 10.1080/19443994.2012.715093.

[480] J. A. Carta, J. González, P. Cabrera, and V. J. Subiela. Preliminary experimental analysis of a small-scale prototype SWRO desalination plant, designed for continuous adjustment of its energy consumption to the widely varying power generated by a stand-alone wind turbine. *Appl. Energy*, 137:222–239, 2015. DOI: 10.1016/j.apenergy.2014.09.093.

[481] T. Bilstad, E. Protasova, A. Simonova, S. Stornes, and I. Yuneizi. Wind-powered RO desalination. *Desalin. Water Treat.*, 55:3106–3110, 2015. DOI: 10.1080/19443994.2014.939873.

[482] F. J. G. Latorre, S. O. P. Báez, and A. G. Gotor. Energy performance of a reverse osmosis desalination plant operating with variable pressure and flow. *Desalination*, 366:146–153, 2015. DOI: 10.1016/j.desal.2015.02.039.

[483] A. M. Bilton, L. C. Kelley, and S. Dubowsky. Photovoltaic reverse osmosis—feasibility and a pathway to develop technology. *Desalin. Water Treat.*, 31:24–34, 2011. DOI: 10.5004/dwt.2011.2398.

[484] D. P. Clarke, Y. M. Al-Abdeli, and G. Kothapalli. The effects of including intricacies in the modelling of a small-scale solar-PV reverse osmosis desalination system. *Desalination*, 311:127–136, 2013. DOI: 10.1016/j.desal.2012.11.006.

[485] L. C. Kelley and S. Dubowsky. Thermal control to maximize photovoltaic powered reverse osmosis desalination systems productivity. *Desalination*, 314:10–19, 2013. DOI: 10.1016/j.desal.2012.11.036.

[486] S. Kumarasamy, S. Narasimhan, and S. Narasimhan. Optimal operation of battery-less solar powered reverse osmosis plant for desalination. *Desalination*, 375:89–99, 2015. DOI: 10.1016/j.desal.2015.07.029.

[487] E. Ntavou, G. Kosmadakis, D. Manolakos, G. Papadakis, and D. Papantonis. Experimental evaluation of a multi-skid reverse osmosis unit operating at fluctuating power input. *Desalination*, 398:77–86, 2016. DOI: 10.1016/j.desal.2016.07.014.

[488] D. Weiner, D. Fisher, E. J. Moses, B. Katz, and G. Meron. Operation experience of a solar- and wind-powered desalination demonstration plant. *Desalination*, 137:7–13, 2001. DOI: 10.1016/s0011-9164(01)00198-9.

[489] S. A. Kershman, J. Rheinländer, and H. Gabler. Seawater reverse osmosis powered from renewable energy sources-hybrid wind/photovoltaic/grid power supply for small-scale desalination in Libya. *Desalination*, 153:17–23, 2003. DOI: 10.1016/s0011-9164(02)01089-5.

[490] A. Hossam-Eldin, A. M. El-Nashar, and A. Ismaiel. Investigation into economical desalination using optimized hybrid renewable energy system. *Int. J. Electr. Power Energy Syst.*, 43:1393–1400, 2012. DOI: 10.1016/j.ijepes.2012.05.019.

[491] E. M. A. Mokheimer, A. Z. Sahin, A. Al-Sharafi, and A. I. Ali. Modeling and optimization of hybrid wind-solar-powered reverse osmosis water desalination system in Saudi Arabia. *Energy Convers. Manag.*, 75:86–97, 2013. DOI: 10.1016/j.enconman.2013.06.002.

[492] H. M. N. AlMadani. Water desalination by solar powered electrodialysis process. *Renew. Energy*, 28:1915–1924, 2003. DOI: 10.1016/s0960-1481(03)00014-4. 135

[493] J. M. Ortiz, E. Expósito, F. Gallud, V. García-García, V. Montiel, and A. Aldaz. Electrodialysis of brackish water powered by photovoltaic energy without batteries: Direct connection behaviour. *Desalination*, 208:89–100, 2007. DOI: 10.1016/j.desal.2006.05.026. 135

[494] M. R. Adiga, S. K. Adhikary, P. K. Narayanan, W. P. Harkare, S. D. Gomkale, and K. P. Govindan. Performance analysis of photovoltaic electrodialysis desalination plant at Tanote in Thar desert. *Desalination*, 67:59–66, 1987. 138

[495] C. Fernandez-Gonzalez, A. Dominguez-Ramos, R. Ibañez, and A. Irabien. Sustainability assessment of electrodialysis powered by photovoltaic solar energy for freshwater production. *Renew. Sustain. Energy Rev.*, 47:604–615, 2015. DOI: 10.1016/j.rser.2015.03.018. 138, 139, 140

[496] M. T. Ali, H. E. S. Fath, and P. R. Armstrong. A comprehensive techno-economical review of indirect solar desalination. *Renew. Sustain. Energy Rev.*, 15:4187–4199, 2011. DOI: 10.1016/j.rser.2011.05.012. 138

[497] N. C. Wright. Justification for community-scale photovoltaic-powered electrodialysis desalination systems for inland rural villages in India. *Desalination*, 352:82–91, 2014. DOI: 10.1016/j.desal.2014.07.035. 138

[498] M. Khayet. Solar desalination by membrane distillation: Dispersion in energy consumption analysis and water production costs (a review). *Desalination*, 308:89–101, 2013. DOI: 10.1016/j.desal.2012.07.010. 140, 142

[499] F. Banat, R. Jumah, and M. Garaibeh. Exploitation of solar energy collected by solar stills for desalination by membrane distillation. *Renew. Energy*, 25:293–305, 2002. DOI: 10.1016/s0960-1481(01)00058-1. 140

[500] R. B. Saffarini, E. K. Summers, and H. A. Arafat. Technical evaluation of stand-alone solar powered membrane distillation systems. *Desalination*, 286:332–341, 2012. DOI: 10.1016/j.desal.2011.11.044. 140

[501] R. Porrazzo, A. Cipollina, M. Galluzzo, and G. Micale. A neural network-based optimizing control system for a seawater-desalination solar-powered membrane distillation unit. *Comput. Chem. Eng.*, 54:79–96, 2013. DOI: 10.1016/j.compchemeng.2013.03.015. 140

[502] F. Banat, N. Jwaied, M. Rommel, J. Koschikowski, and M. Wieghaus. Desalination by a compact SMADES autonomous solar-powered membrane distillation unit. *Desalination*, 217:29–37, 2007. DOI: 10.1016/j.desal.2006.11.028. 141

[503] X. Wang, L. Zhang, H. Yang, and H. Chen. Feasibility research of potable water production via solar-heated hollow fiber membrane distillation system. *Desalination*, 247:403–411, 2009. DOI: 10.1016/j.desal.2008.10.008. 142

[504] J.-P. Mericq, S. Laborie, and C. Cabassud. Evaluation of systems coupling vacuum membrane distillation and solar energy for seawater desalination. *Chem. Eng. J.*, 166:596–606, 2011. DOI: 10.1016/j.cej.2010.11.030. 142

[505] T.-C. Chen and C.-D. Ho. Immediate assisted solar direct contact membrane distillation in saline water desalination. *J. Memb. Sci.*, 358:122–130, 2010. DOI: 10.1016/j.memsci.2010.04.037. 142

[506] R. B. Saffarini, E. K. Summers, and H. A. Arafat. Economic evaluation of stand-alone solar powered membrane distillation systems. *Desalination*, 299:55–62, 2012. DOI: 10.1016/j.desal.2012.05.017.

[507] D. Winter, J. Koschikowski, and S. Ripperger. Desalination using membrane distillation: Flux enhancement by feed water deaeration on spiral-wound modules. *J. Memb. Sci.*, 423:215–224, 2012. DOI: 10.1016/j.memsci.2012.08.018.

[508] F. Suárez and R. Urtubia. Tackling the water-energy nexus: An assessment of membrane distillation driven by salt-gradient solar ponds. *Clean Technol. Environ. Policy*, 18:1697–1712, 2016. DOI: 10.1007/s10098-016-1210-3.

[509] A. Cipollina, M. G. Di Sparti, A. Tamburini, and G. Micale. Development of a membrane distillation module for solar energy seawater desalination. *Chem. Eng. Res. Des.*, 90:2101–2121, 2012. DOI: 10.1016/j.cherd.2012.05.021.

[510] C. L. Ong, W. Escher, S. Paredes, A. S. G. Khalil, and B. Michel. A novel concept of energy reuse from high concentration photovoltaic thermal (HCPVT) system for desalination. *Desalination*, 295:70–81, 2012. DOI: 10.1016/j.desal.2012.04.005.

[511] E. Guillén-Burrieza, G. Zaragoza, S. Miralles-Cuevas, and J. Blanco. Experimental evaluation of two pilot-scale membrane distillation modules used for solar desalination. *J. Memb. Sci.*, 409:264–275, 2012. DOI: 10.1016/j.memsci.2012.03.063.

[512] H. Chang, S.-G. Lyu, C.-M. Tsai, Y.-H. Chen, T.-W. Cheng, and Y.-H. Chou. Experimental and simulation study of a solar thermal driven membrane distillation desalination process. *Desalination*, 286:400–411, 2012. DOI: 10.1016/j.desal.2011.11.057.

[513] Y. Wang, Z. Xu, N. Lior, and H. Zeng. An experimental study of solar thermal vacuum membrane distillation desalination. *Desalin. Water Treat.*, 53:887–897, 2015. DOI: 10.1080/19443994.2014.927187.

[514] X. Li, Y. Qin, R. Liu, Y. Zhang, and K. Yao. Study on concentration of aqueous sulfuric acid solution by multiple-effect membrane distillation. *Desalination*, 307:34–41, 2012. DOI: 10.1016/j.desal.2012.08.023.

[515] F. Banat and N. Jwaied. Autonomous membrane distillation pilot plant unit driven solar energy: Experiences and lessons learned. *Int. J. Sustain. Water Environ. Syst.*, 1:21–24, 2010. DOI: 10.5383/swes.0101.005.

[516] R. G. Raluy, R. Schwantes, V. J. Subiela, B. Peñate, G. Melián, and J. R. Betancort. Operational experience of a solar membrane distillation demonstration plant in Pozo Izquierdo-Gran Canaria Island (Spain). *Desalination*, 290:1–13, 2012. DOI: 10.1016/j.desal.2012.01.003.

[517] A. Chafidz, S. Al-Zahrani, M. N. Al-Otaibi, C. F. Hoong, T. F. Lai, and M. Prabu. Portable and integrated solar-driven desalination system using membrane distillation for arid remote areas in Saudi Arabia. *Desalination*, 345:36–49, 2014. DOI: 10.1016/j.desal.2014.04.017.

[518] K. Nakoa, K. Rahaoui, A. Date, and A. Akbarzadeh. An experimental review on coupling of solar pond with membrane distillation. *Sol. Energy*, 119:319–331, 2015. DOI: 10.1016/j.solener.2015.06.010.

[519] G. Zaragoza, A. Ruiz-Aguirre, and E. Guillén-Burrieza. Efficiency in the use of solar thermal energy of small membrane desalination systems for decentralized water production. *Appl. Energy*, 130:491–499, 2014. DOI: 10.1016/j.apenergy.2014.02.024.

[520] F. Suárez and S. W. Tyler. Comments on Evaluation of systems coupling vacuum membrane distillation and solar energy for seawater desalination. *Chem. Eng. J.*, pages 475–476, 2011.

[521] F. Suárez, J. A. Ruskowitz, S. W. Tyler, and A. E. Childress. Renewable water: Direct contact membrane distillation coupled with solar ponds. *Appl. Energy*, 158:532–539, 2015.

[522] A. Ruiz-Aguirre, M. I. Polo-López, P. Fernández-Ibáñez, and G. Zaragoza. Assessing the validity of solar membrane distillation for disinfection of contaminated water. *Desalin. Water Treat.*, 55:2792–2799, 2015. DOI: 10.1080/19443994.2014.946717.

[523] R. Schwantes, A. Cipollina, F. Gross, J. Koschikowski, D. Pfeifle, M. Rolletschek, and V. Subiela. Membrane distillation: Solar and waste heat driven demonstration plants for desalination. *Desalination*, 323:93–106, 2013. DOI: 10.1016/j.desal.2013.04.011.

[524] C.-D. Ho, C. A. Ng, P.-H. Wang, and C.-H. Cheng. Theoretical and experimental studies of immediate assisted solar air gap membrane distillation systems. *Desalin. Water Treat.*, 57:3846–3860, 2016. DOI: 10.1080/19443994.2014.989274.

Authors' Biographies

ERIC M.V. HOEK

Dr. Eric M.V. Hoek is a professor in UCLA's Department of Civil & Environmental Engineering, Institute of the Environment & Sustainability and the California NanoSystems Institute. He is also the Director of the *UCLA Sustainable LA Grand Challenge*. His research explores the union of membrane technologies, nanomaterials and electrochemistry for water, energy and environmental applications. He has over 200 technical publications including over 70 patents filed globally. He has also co-founded several technology startups and has advised a wide array of state, federal and international government agencies, local water utilities, technology companies, investment funds, law firms and research funding agencies.

DAVID JASSBY

David Jassby is an associate professor in the Department of Civil and Environmental Engineering at UCLA. He received his Ph.D. in Civil and Environmental Engineering from Duke University (2011), an M.S. in Civil and Environmental Engineering from UC Davis (2005), and a B.Sc. in Biology from Hebrew University (2002). David spent a year working as a consultant in an Environmental Engineering consulting firm in NC (BBL, Inc.) David's research is primarily concerned with membrane separations, environmental electrochemistry, and water treatment technologies. His lab is currently engaged in research concerning membrane development, desalination, industrial wastewater treatment, oil/water separations, and the electrochemical treatment of contaminated water. He holds several patents on electroactive membranes and processes, and has published more than 60 peer-reviewed manuscripts in peer-reviewed journals.

RICHARD B. KANER

Richard B. Kaner is a Distinguished Professor in the UCLA Departments of Chemistry and Materials Science and Engineering and holds the Dr. Hong Endowed Chair in Materials Innovation. His research explores new materials from graphene for energy storage to nanostructured conducting polymers for separation applications. He has published over 425 papers and 45 U.S. patents. According to Clarivate Analytics and Thomson–Reuters he is among the world's most highly cited authors. He has received Fellowships from Dreyfus, Fulbright, Guggenheim, Packard, and Sloan Foundations along with the Materials Research Society Medal, the Royal Society of Chemistry Centenary Prize, the Chemical Pioneer Award, and the American Chemical Society Award in the Chemistry of Materials.

JISHAN WU

Jishan Wu is a Ph.D. student in UCLA's Department of Civil & Environmental Engineering, under the supervision of Prof. Eric M.V. Hoek. His research revolves around membrane and other advanced technologies for water applications. He is the lead author of the book Sustainable Desalination and Water Reuse. His Ph.D. thesis focuses on developing novel ultra-high pressure reverse osmosis (UHPRO) with the objective of achieving minimum/zero liquid discharge more cost-effective and sustainable.

JINGBO WANG

Dr. Jingbo Wang is currently a postdoctoral scholar working with Prof. Eric Hoek and Prof. David Jassby in the Department of Civil & Environmental Engineering at UCLA. She received her B.S. from Wuhan University in Water and Wastewater Engineering, and Ph.D. from University of North Carolina at Chapel Hill in Environmental Sciences and Engineering. Her research interests focus on desalination, water and wastewater treatment with membrane technologies, and application of nanotechnology in membrane separation and other physical/chemical processes in environmental systems.

YIMING LIU

Yiming Liu is currently a Ph.D. student in the Department of Civil and Environmental Engineering at UCLA. His research with Prof. Eric Hoek is modeling membrane-based desalination processes that enables high water recovery. Yiming obtained his Master's degree in Civil Engineering from UCLA and his Bachelor's degree in Environmental Engineering from Tsinghua University.

UNNATI RAO

Unnati Rao is a Postdoctoral Scholar in the department of Civil and Environmental Engineering at the University of California, Los Angeles (UCLA) in the NANOMETER lab under the guidance of Dr. Eric M.V. Hoek. Her research focusses on the development and testing of membrane materials for water treatment. She has also extensively worked in the application of electrochemistry for water treatment processes. She has over six technical publications. She also has a patent and has presented her work in two national level conferences.

Printed in the United States
by Baker & Taylor Publisher Services